Elegant Chaos

Algebraically Simple Chaotic Flows

Elegant Chaos

Algebraically Simple Chaotic Flows

Julien Clinton Sprott
University of Wisconsin-Madison, USA

 **World Scientific**

NEW JERSEY · LONDON · SINGAPORE · BEIJING · SHANGHAI · HONG KONG · TAIPEI · CHENNAI

Published by

World Scientific Publishing Co. Pte. Ltd.

5 Toh Tuck Link, Singapore 596224

USA office: 27 Warren Street, Suite 401-402, Hackensack, NJ 07601

UK office: 57 Shelton Street, Covent Garden, London WC2H 9HE

British Library Cataloguing-in-Publication Data
A catalogue record for this book is available from the British Library.

First published 2010 (Hardcover)
Reprinted 2016 (in paperback edition)
ISBN 978-981-3203-54-9

ELEGANT CHAOS
Algebraically Simple Chaotic Flows

ISBN-13 978-981-283-881-0
ISBN-10 981-283-881-3

Printed in Singapore

Dedicated to the memory of
Edward Norton Lorenz

Edward Norton Lorenz, 1917–2008
Photo: Massachusetts Institute of Technology

Preface

When Ed Lorenz discovered chaos in a simple system of ordinary differential equations in 1959, he unleashed on the world a new field of science that has grown ever larger with each passing year. The fact that simple equations can have solutions of incredible complexity continues to enthrall scientists and raises the hope that phenomena previously thought too complicated to be understood might be adequately described by very simple models. By now, many such models have been developed and studied in great detail, but they continue to present surprises and raise questions, not the least of which is why they had not already been widely known.

From the start of my interest in the field some twenty years ago, I have been intrigued by the quest for the mathematically simplest systems of various types that can exhibit chaos. I was proud to discover some new systems that are in some sense simpler than those previously known or are otherwise more 'elegant' by virtue of the number of parameters or their values or of some special symmetry or economy of notation. This book is an attempt to share those discoveries and to catalog other simple examples that were previously known or recently discovered by others, as well as many cases that are published here for the first time and are thus ripe for further study.

I have also included a chapter at the end on chaotic electrical circuits since the quest for simple chaotic circuits is closely related to the quest for simple equations that exhibit chaos. Some of these circuits are new and provide an opportunity for study and exploration if your inclinations are more toward building things than sitting in front of a computer.

This book should be of interest to chaos researchers looking for simple systems and circuits to use in their studies or for further exploration, to instructors who want examples to teach and motivate students, and to students doing independent study. The book assumes only an elementary knowledge of calculus. The systems are initial-value ordinary differential equations (ODEs), as well as some partial differential equations (PDEs) and delay differential equations (DDEs), but they must be solved numerically, and so a formal course in differential equations is of limited use.

You will get the most out of this book if you can write simple computer programs in the language of your choice or have access to software that allows you to solve systems of coupled ODEs and to display the results graphically. All the calculations and figures in this book were done with the PowerBASIC console compiler (http://www.powerbasic.com/), which I highly recommend. There is no substitute for the thrill and insight of seeing the solution of a simple equation unfold as the trajectory wanders in real time across your computer screen using a program of your own making. A goal of this book is to inspire and delight as well as to teach. I hope you will enjoy reading and studying it as much as I did writing it.

Many people have contributed to the ideas in this book. My greatest debt is to George Rowlands who lured me away from plasma physics when he got me interested in chaos in the late 1980s and who continues to be a valued colleague and mentor. I am also grateful to other collaborators with whom I have coauthored papers on these topics including Cliff Pickover, Dee Dechert, Stefan Linz, Wendell Horton, Karl Lonngren, Hans Gottlieb, Ken Kiers, Wajdi Ahmad, Jim Crutchfield, John Vano, Konstantinos Chlouverakis, Zeraoulia Elhadj, Kehui Sun, and Del Marshall. Working with such interesting and talented people is one of my greatest delights. I am also thankful to the students with whom I have worked over the years, especially those who were able to publish their results including Christopher Watts, David Newman, Brian Meloon, Kevin Mirus, David Albers, Joe Wildenberg, Mike Anderson, Jeff Noel, Joseph Azizi, Charles Brummitt, Adam Maus and Vladimir Zhdankin. They have challenged me with difficult questions and helped study some of the systems in this book. Finally, I would like to thank Jessica Piper for proofreading an early version of the manuscript and for constructing and testing some of the new chaotic circuits in the final chapter.

J. C. Sprott
Madison, Wisconsin
December 2009

Contents

Preface   vii

List of Tables   xv

1. Fundamentals   1

    1.1   Dynamical Systems   1
    1.2   State Space   2
    1.3   Dissipation   7
    1.4   Limit Cycles   8
    1.5   Chaos and Strange Attractors   10
    1.6   Poincaré Sections and Fractals   12
    1.7   Conservative Chaos   16
    1.8   Two-toruses and Quasiperiodicity   18
    1.9   Largest Lyapunov Exponent   20
    1.10   Lyapunov Exponent Spectrum   24
    1.11   Attractor Dimension   29
    1.12   Chaotic Transients   31
    1.13   Intermittency   32
    1.14   Basins of Attraction   32
    1.15   Numerical Methods   36
    1.16   Elegance   37

2. Periodically Forced Systems   41

    2.1   Van der Pol Oscillator   41
    2.2   Rayleigh Oscillator   43
    2.3   Rayleigh Oscillator Variant   43
    2.4   Duffing Oscillator   44

2.5   Quadratic Oscillators   47
2.6   Piecewise-linear Oscillators   48
2.7   Signum Oscillators   49
2.8   Exponential Oscillators   51
2.9   Other Undamped Oscillators   51
2.10   Velocity Forced Oscillators   53
2.11   Parametric Oscillators   55
2.12   Complex Oscillators   57

3. Autonomous Dissipative Systems   61

3.1   Lorenz System   61
3.2   Diffusionless Lorenz System   64
3.3   Rössler System   66
3.4   Other Quadratic Systems   68
       3.4.1   Rössler prototype-4 system   68
       3.4.2   Sprott systems   68
3.5   Jerk Systems   70
       3.5.1   Simplest quadratic case   73
       3.5.2   Rational jerks   76
       3.5.3   Cubic cases   77
       3.5.4   Cases with arbitrary power   79
       3.5.5   Piecewise-linear case   80
       3.5.6   Memory oscillators   82
3.6   Circulant Systems   83
       3.6.1   Halvorsen's system   84
       3.6.2   Thomas' systems   85
       3.6.3   Piecewise-linear system   86
3.7   Other Systems   86
       3.7.1   Multiscroll systems   87
       3.7.2   Lotka–Volterra systems   88
       3.7.3   Chua's systems   90
       3.7.4   Rikitake dynamo   92

4. Autonomous Conservative Systems   95

4.1   Nosé–Hoover Oscillator   95
4.2   Nosé–Hoover Variants   97
4.3   Jerk Systems   98
       4.3.1   Jerk form of the Nosé–Hoover oscillator   98

4.3.2 Simplest conservative chaotic flow 99

4.3.3 Other conservative jerk systems 99

4.4 Circulant Systems 101

4.4.1 Quadratic case 102

4.4.2 Cubic case 102

4.4.3 Labyrinth chaos 105

4.4.4 Piecewise-linear system 107

5. Low-dimensional Systems ($D < 3$) 109

5.1 Dixon System 109

5.2 Dixon Variants 110

5.3 Logarithmic Case 112

5.4 Other Cases 114

6. High-dimensional Systems ($D > 3$) 115

6.1 Periodically Forced Systems 115

6.1.1 Forced pendulum 116

6.1.2 Other forced nonlinear oscillators 118

6.2 Master–slave Oscillators 118

6.3 Mutually Coupled Nonlinear Oscillators 120

6.3.1 Coupled pendulums 121

6.3.2 Coupled van der Pol oscillators 123

6.3.3 Coupled FitzHugh–Nagumo oscillators 123

6.3.4 Coupled complex oscillators 124

6.3.5 Other coupled nonlinear oscillators 125

6.4 Hamiltonian Systems 126

6.4.1 Coupled nonlinear oscillators 128

6.4.2 Velocity coupled oscillators 129

6.4.3 Parametrically coupled oscillators 130

6.4.4 Simplest Hamiltonian 130

6.4.5 Hénon–Heiles system 132

6.4.6 Reduced Hénon–Heiles system 133

6.4.7 N-body gravitational systems 134

6.4.8 N-body Coulomb systems 138

6.5 Anti-Newtonian Systems 142

6.5.1 Two-body problem 142

6.5.2 Three-body problem 145

6.6 Hyperjerk Systems 147

	6.6.1	Forced oscillators	147
	6.6.2	Chlouverakis systems	148
6.7	Hyperchaotic Systems		152
	6.7.1	Rössler hyperchaos	153
	6.7.2	Snap hyperchaos	154
	6.7.3	Coupled chaotic systems	154
	6.7.4	Other hyperchaotic systems	156
6.8	Autonomous Complex Systems		156
6.9	Lotka–Volterra Systems		157
6.10	Artificial Neural Networks		159
	6.10.1	Minimal dissipative artificial neural network	161
	6.10.2	Minimal conservative artificial neural network	162
	6.10.3	Minimal circulant artificial neural network	162

7. **Circulant Systems** — 165

7.1	Lorenz–Emanuel System		165
7.2	Lotka–Volterra Systems		169
7.3	Antisymmetric Quadratic System		171
7.4	Quadratic Ring System		171
7.5	Cubic Ring System		171
7.6	Hyperlabyrinth System		173
7.7	Circulant Neural Networks		174
7.8	Hyperviscous Ring		176
7.9	Rings of Oscillators		176
	7.9.1	Coupled pendulums	177
	7.9.2	Coupled cubic oscillators	177
	7.9.3	Coupled signum oscillators	178
	7.9.4	Coupled van der Pol oscillators	179
	7.9.5	Coupled FitzHugh–Nagumo oscillators	180
	7.9.6	Coupled complex oscillators	182
	7.9.7	Coupled Lorenz systems	182
	7.9.8	Coupled jerk systems	185
7.10	Star Systems		185
	7.10.1	Coupled pendulums	187
	7.10.2	Coupled cubic oscillators	187
	7.10.3	Coupled signum oscillators	188
	7.10.4	Coupled van der Pol oscillators	190
	7.10.5	Coupled FitzHugh–Nagumo oscillators	191
	7.10.6	Coupled complex oscillators	191

7.10.7 Coupled diffusionless Lorenz systems 193
7.10.8 Coupled jerk systems 194

8. Spatiotemporal Systems 195

8.1 Numerical Methods 195
8.2 Kuramoto–Sivashinsky Equation 199
8.3 Kuramoto–Sivashinsky Variants 200
    8.3.1 Cubic case 201
    8.3.2 Quartic case 201
8.4 Chaotic Traveling Waves 201
    8.4.1 Rotating Kuramoto–Sivashinsky system 203
    8.4.2 Rotating Kuramoto–Sivashinsky variant 203
8.5 Continuum Ring Systems 204
    8.5.1 Quadratic ring system 204
    8.5.2 Antisymmetric quadratic system 205
    8.5.3 Other simple PDEs 207
8.6 Traveling Wave Variants 212

9. Time-Delay Systems 221

9.1 Delay Differential Equations 221
9.2 Mackey–Glass Equation 223
9.3 Ikeda DDE 223
9.4 Sinusoidal DDE 225
9.5 Polynomial DDE 225
9.6 Sigmoidal DDE 227
9.7 Signum DDE 227
9.8 Piecewise-linear DDEs 229
    9.8.1 Antisymmetric case 229
    9.8.2 Asymmetric case 229
    9.8.3 Asymmetric logistic DDE 230
9.9 Asymmetric Logistic DDE with Continuous Delay 232

10. Chaotic Electrical Circuits 233

10.1 Circuit Elegance 233
10.2 Forced Relaxation Oscillator 234
10.3 Autonomous Relaxation Oscillator 237
10.4 Coupled Relaxation Oscillators 239
    10.4.1 Two oscillators 239
    10.4.2 Many oscillators 241

10.5   Forced Diode Resonator   242

10.6   Saturating Inductor Circuit   243

10.7   Forced Piecewise-linear Circuit   246

10.8   Chua's Circuit   246

10.9   Nishio's Circuit   249

10.10   Wien-bridge Oscillator   251

10.11   Jerk Circuits   254

        10.11.1 Absolute-value case   254

        10.11.2 Single-knee case   255

        10.11.3 Signum case   256

        10.11.4 Signum variant   258

10.12   Master–slave Oscillator   259

10.13   Ring of Oscillators   261

10.14   Delay-line Oscillator   263

Bibliography   265

Index   281

List of Tables

1.1   Characteristics of the attractors for a bounded three-dimensional flow 28

2.1   Chaotic forced damped quadratic oscillators 48

2.2   Chaotic forced conservative nonlinear oscillators 52

2.3   Chaotic velocity forced oscillators from Eq. (2.15) with $A = 1$ and $\Omega = 1$ 53

2.4   Chaotic nonlinear parametric oscillators 55

2.5   Chaotic forced complex oscillators 57

3.1   Simple three-dimensional chaotic flows with quadratic nonlinearities 70

3.2   Simple chaotic jerk systems 72

3.3   Chaotic memory oscillators 74

3.4   Variants of Chua's circuit with $\dot{y} = x + z$ and $\dot{z} = -y$ 91

4.1   Chaotic conservative jerk systems 101

5.1   Simplified variants of the Dixon system 111

6.1   Forced nonlinear oscillators with $\dot{y} = u$ and $\dot{u} = -\Omega^2 y$ 118

6.2   Chaotic master–slave oscillators 120

6.3   Chaotic coupled nonlinear oscillators 126

6.4   Chaotic Chlouverakis snap systems 149

6.5   Chaotic autonomous complex systems 157

8.1   Chaotic partial differential equations 207

8.2   Chaotic traveling wave PDE variants 213

Chapter 1

Fundamentals

This chapter will describe the fundamental concepts of continuous-time chaotic systems so as to make the book self-contained and readable by someone unfamiliar with the subject using the damped, forced pendulum as the main example. It will also define the sense in which a chaotic system is deemed to be elegant since that term is rather subjective with no universal definition. Most of the material in this chapter is covered in more detail in Sprott (2003).

1.1 Dynamical Systems

A *dynamical system* is one whose state changes in time. If the changes are determined by specific rules, rather than being random, we say that the system is *deterministic*; otherwise it is *stochastic*. The changes can occur at discrete time steps or continuously. This book will be concerned with continuous-time, deterministic, dynamical systems since they arguably best approximate the real world. This view represents the prejudice of most physical scientists, but it is also the case that chaos is relatively too easy to achieve in discrete-time systems, and hence it is less of a challenge to find elegant examples of chaos in such systems, and those that are found have less apparent relevance to the natural world. Also, discrete-time systems have already been extensively explored, in part because they are more computationally tractable.

Stochastic systems mimic many of the features of chaos, but they are not chaotic because chaos is a property of deterministic systems. Furthermore, introducing randomness into a dynamical model is a way of admitting ignorance of the underlying process and obtaining plausible behavior without a deep understanding of its cause. For these reasons, stochastic systems will not be discussed in this book.

The most obvious examples of dynamical systems are those that involve something moving through space, like a planet orbiting the Sun, a pendulum swinging back and forth, or an animal exploring its habitat. But dynamical systems can be more abstract, such as money flowing through the economy, information propagating across the Internet, or disease spreading through the population. In each of these examples, we characterize the state of the system at each instant by a set of values of the time-changing *variables* that collectively define not only the current conditions but that uniquely determine what will happen in the future.

For a planet orbiting the Sun, six variables are required, three components to describe its position relative to the Sun at each instant, and three to describe its velocity in the three-dimensional space in which it moves. Newton's second law ($F = ma$) coupled with the universal law of gravitation ($F = GmM/R^2$) provides the deterministic rule whereby its future state is completely and uniquely determined.

Calculation of celestial motion is one of earliest problems solved by scientists, and the calculations are some the most precise in all of science, allowing, for example, the prediction of Solar eclipses many years in advance both in time and space. Not only can one predict when such an eclipse will occur, but where on the Earth to stand to get the best view of it. Contrast this strong predictability with the difficulty of predicting even a few days in advance whether the sky will be clear enough to observe that eclipse using extremely detailed weather models and vast computational resources. Therein lies the difference between regular and chaotic dynamics.

For the simpler example of a pendulum swinging in a plane (Baker and Gollub, 1996; Baker and Blackburn, 2005), only two variables are required, one to specify its angle with respect to the vertical, and a second to specify its velocity. Such a system has one spatial degree of freedom (the arc of a circle), in contrast to the planet orbiting the Sun, which has three degrees of freedom, and this one degree of freedom is more than sufficient to produce chaos under some conditions. Hence it is a simpler and more elegant example with which to begin.

1.2   State Space

Consider a 1-kilogram mass on the end of a 1-meter-long, massless, rigid rod (so as to allow oscillations of arbitrarily large amplitude, including ones where the pendulum goes 'over the top') as shown in Fig. 1.1, and located on

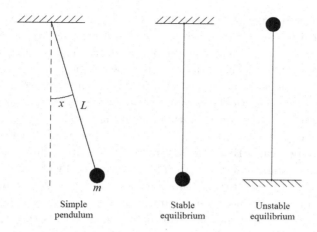

Simple pendulum   Stable equilibrium   Unstable equilibrium

Fig. 1.1   A pendulum consisting of a mass m on the end of massless, rigid rod of length L at an angle (in radians) of x. The quantities m and L are fixed parameters of the system, and x is a time-dependent variable with $x = 0$ corresponding to a stable equilibrium and $x = \pi$ corresponding to an unstable equilibrium.

a distant planet where the acceleration due to gravity is 1 meter/second^2 (about one tenth of Earth's gravity). These particular values make the resulting equations more elegant in a sense to be described later and adequately serve to illustrate the ideas, which are very general. Of course such an idealized pendulum could never be built, not only because massless rods are unavailable, but because any rod with a mass much less than 1 kilogram would not likely have sufficient strength to withstand the buckling forces when the pendulum is inverted, but these practical limitations need not constrain our imagination or deter the discussion.

The mass, length, and acceleration due to gravity are considered *parameters* of the system that can be adjusted but that do not change during the time the motion is being examined. A major theme of this book is to find values of the parameters that make the equations simple but that allow the system to behave chaotically. In the equations that model the system, the parameters will usually be denoted by letters near the beginning of the alphabet $(a, A, b, \ldots)$ or sometimes by Greek letters $(\alpha, \beta, \gamma, \Omega, \sigma, \ldots)$, whereas the variables will be denoted by letters near the end of the alphabet (t, u, v, w, x, y, z). For a dynamical system, the time (here always denoted by t) is the *independent variable* upon which the other *dependent variables* depend.

The distinction between parameters and variables is usually clear in a mathematical model, but is much less so in a real dynamical system. In a radio receiver, the parameters might be the position of the frequency and volume knobs, whereas the variables might be the position and velocity of the cone of the loudspeaker as it vibrates in response to the audio signal that is received for particular settings of the knobs. In a climate model, a parameter might be the concentration of carbon dioxide in the atmosphere, or perhaps its rate of change, whereas a variable might be the globally averaged temperature. However, in a more complicated climate model, the carbon dioxide might itself be a variable determined by the model.

Think of the parameters as time-independent inputs to the model that are specified by factors not included in the model and the variables as time-dependent outputs that are determined by the equations that specify the model. In the real world, almost everything is a variable expect perhaps a few fundamental constants such as the speed of light and Planck's constant, but the systems in this book are gross simplifications of nature in which only a few quantities are allowed to vary while others are held rigidly constant. The elegance comes from showing that interesting and realistic behavior can result from such extremely simple models.

In the pendulum swinging in a plane, one time-dependent variable is the angle x (in radians) that the pendulum makes from the vertical. At each instant, there is a force equal to $-\sin x$ pulling the mass back to its stable equilibrium position at $x = 0$. The virtue of using the parameters $m = L = g = 1$ is that the coefficient multiplying $\sin x$, which would normally be g/L, is eliminated. Newton's second law then leads to a system of equations for the motion given by

$$\dot{x} = v$$
$$\dot{v} = -\sin x,$$

(1.1)

where v is the angular velocity and the overdot denotes a time derivative ($\dot{x} = dx/dt$). (Isaac Newton used this notation when he invented calculus, whereas his rival, Gottfried Leibniz, used the more familiar but more clumsy notation.) This is the sense in which a mechanical system with one degree of freedom has a *state space* of two dimensions (x and v). A state space such as this, in which one variable is proportional to the time derivative of another ($v = dx/dt$), is usually called *phase space*, but we will keep the more general term throughout this book since the state space variables are not always related in this way.

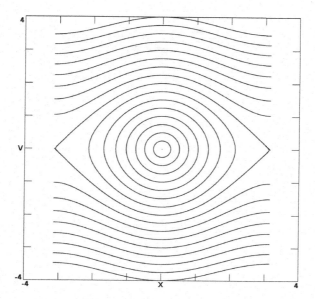

Fig. 1.2 State space plot for the frictionless pendulum from Eq. (1.1) with many different initial conditions showing representative streamlines of the flow. The closed curves near the origin represent librations, and the open lines at large values of $|v|$ represent rotations.

Equations (1.1) imply that at every point in this abstract two-dimensional state space of x and v, there is a unique direction and amplitude of the motion given by the vector whose components are $\dot{x}$ and $\dot{v}$. The simultaneous motion of all the points in this space resembles a flowing fluid, and hence systems such as Eq. (1.1) are usually called *flows*. Figure 1.2 shows some of the streamlines for this particular flow. The streamlines are simply the clockwise trajectories followed by small particles moving with the flow for various initial conditions. In this case, each streamline is a curve of constant total energy (kinetic + potential) given by $E = v^2/2 + 1 - \cos x$, with the different curves corresponding to different values of E. Of course, there is a streamline through every point in the state space, only a representative sample of which is shown in the figure.

There are several things to note about Fig. 1.2. There are two kinds of streamlines — those that encircle the origin at $x = v = 0$, and those that bump along at large positive or negative velocities. The former represent back and forth oscillations of the pendulum (so-called *librations*), and the

Fig. 1.3   State space plot for the frictionless pendulum from Eq. (1.1) with many different initial conditions showing representative streamlines of the flow as in Fig. 1.2 but wrapped onto a cylinder and then projected back onto the plane of the page.

latter represent *rotations* (sometimes called 'hindered rotations' since gravity affects the rotational velocity) in which the pendulum goes 'over the top.' For rotational motion, the angle continually increases or decreases, and the streamlines are plotted mod 2π so that when they reach $+\pi$ they wrap around and continue at $-\pi$ but going in the same direction, and similarly when they reach $-\pi$. Curiously, the cases are counterintuitive since the streamlines that are closed loops correspond to back-and-forth oscillations, while those that appear to end at $+\pi$ and $-\pi$ are rotations.

If the streamlines were drawn on the surface of a cylinder of unit radius and circumference 2π, they would automatically wrap around in the correct way as Fig. 1.3 shows. Of course, the cylinder is projected back onto the surface of the page, and so a bit of imagination is required to see how the trajectories wrap around it. The visualization is aided by making the lines darkest where they are closest to the viewer and by showing a subtle shadow as if the cylinder were illuminated by a light coming from the top left of the figure.

The two types of streamlines are separated by a line called the *separatrix* that appears to have a sharp bend in Fig. 1.2. That bend is called an

X-point since it appears to cross itself when drawn on the surface of a cylinder as in Fig. 1.3, and it corresponds to the pendulum swinging with just enough energy ($E = 2$) to reach the inverted position with zero velocity. In such a case, it would stall and remain there forever. The X-point and the *O-point* at the origin (also called a *center*) are equilibrium points for the pendulum, with the former being unstable since a slight perturbation will grow, while the latter is stable since a small perturbation will simply cause a small rotation around the center. These two equilibrium conditions are shown in Fig. 1.1.

The X-point is also called a *saddle point* since, like the saddle on a horse, there are two directions (the *stable manifold*) from which a ball rolling downhill will approach the equilibrium and two other directions (the *unstable manifold*) toward which a ball released from near the equilibrium point will roll downhill. Potato chips also sometimes have this shape. The analogy is not perfect, however, since the velocity of the flow goes to zero at the equilibrium point, whereas a ball rolling along the stable manifold toward the equilibrium point of a real saddle will shoot right past it without slowing down. Saddle points also exist in higher-dimensional state spaces where the manifolds have a correspondingly higher dimension and are thus harder to visualize.

1.3 Dissipation

In the previous example, there was no friction, and hence the mechanical energy is conserved and the pendulum swings or rotates forever. The streamlines on a cylinder are closed curves. Such systems are said to be *conservative* since they conserve energy. According to Liouville's theorem (Lichtenberg and Lieberman, 1992), they also have the property that the area occupied by a cluster of initial conditions remains constant in time, implying that the flow is incompressible, much like a swirling liquid. The flow is also time-reversible since changing the sign of t corresponds to the transformation $v \rightarrow -v$, which flips the figure vertically and preserves the shape of the streamlines. Strictly conservative systems are rare in nature, with the standard examples being the motion of astronomical bodies in the near vacuum of outer space and the nonrelativistic motion of individual charged particles in magnetic fields where electromagnetic radiation can be neglected.

For the pendulum, it is more realistic to include a friction term in the equations of motion, due primarily to air resistance (or whatever gas exists in the atmosphere of the distant planet). This friction term is typically assumed to be proportional to the velocity, which is reasonable approximation if the velocity is not too large, and directed opposite to it, so that the equations of motion become

$$\dot{x} = v$$
$$\dot{v} = -bv - \sin x,$$

(1.2)

where b is a measure of the friction. The quantity b is a parameter that is assumed to have a constant value as the motion continues, but that can be changed to produce different types of motion. More generally, such a term is called *damping*, and it can be a nonlinear function of v and can also depend on x.

Systems with damping are said to be *dissipative*, in contrast to the conservative system previously discussed. Dissipative systems do not conserve mechanical energy and are not time-reversible. Their streamlines are not contours of constant energy. They are described by a *compressible* flow as shown in Fig. 1.4 for a case with a small damping of $b = 0.05$ and an initial condition of $(x_0, v_0) = (0, 4)$. In this case, the pendulum rotates eight times and then swings back and forth with a decreasing amplitude, approaching ever closer to the stable equilibrium point at $(x, v) = (0, 0)$. This point acts as an *attractor* for all initial conditions since they are drawn to it in the limit of $t \to \infty$, and it is called a *sink* for obvious reasons. Imagine the fluid swirling as it goes down the drain. Hence instead of just saying 'the pendulum slows down,' we can now say, in the pompous language of dynamical systems, that 'all initial conditions in state space are attracted to the stable equilibrium at the origin.'

1.4   Limit Cycles

The equilibrium at the origin in the previous example is a point attractor for $b > 0$, but other types of attractors also exist. Consider a situation in which the damping is negative ($b < 0$) near the origin but then becomes positive when the trajectory gets too far from the origin. In such a case, the trajectory is drawn to the region where the two effects offset. A system with this property is

$$\dot{x} = v$$
$$\dot{v} = (1 - x^2)v - \sin x,$$

(1.3)

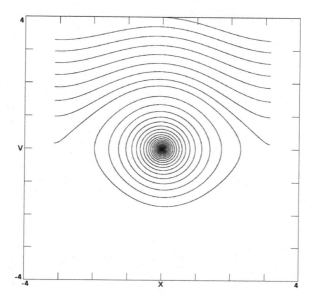

Fig. 1.4 State space plot showing a point attractor for the damped pendulum in Eq. (1.2) with $b = 0.05$ for an initial condition of $(x_0, v_0) = (0, 4)$.

where a coefficient of unity has been chosen for the damping, which causes rapid convergence to a kind of attractor called a *limit cycle*. Figure 1.5 shows the trajectories for eight different initial conditions, all of which approach the attractor, as do almost all initial conditions in the state space. The exceptions are the O-point and the X-point, along with its stable manifold. The O-point is still an equilibrium, but it is now unstable and is called a *repellor* rather than an attractor. It is sometimes called a *source* for obvious reasons. Similar behavior also occurs in a simpler system called the *van der Pol oscillator* (van der Pol, 1920, 1926) where the $\sin x$ in Eq. (1.3) is replaced by x, which eliminates the X-point.

Note that attractors in flows can be zero-dimensional (an equilibrium point) or one-dimensional (a limit cycle). Note also that the curves in Fig. 1.2 are not limit cycles, but rather are *invariant circles* since they do not attract nearby initial conditions. They are circles only in the topological sense of being closed loops and can be greatly distorted in shape. Attractors are a feature of dissipative systems and do not occur in conservative systems.

The antidamping that occurs near the origin implies an external source of energy, not explicit in the equations, and corresponds to positive feed-

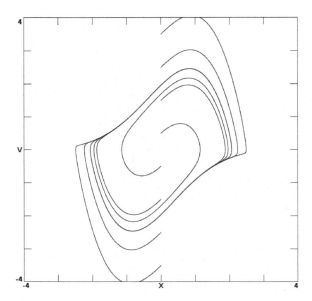

Fig. 1.5 State space plot showing a limit cycle for the damped pendulum in Eq. (1.3) for various initial conditions.

back. In fact, a public address system where the microphone gets too close to the speaker is an excellent example of a limit cycle, producing an unpleasant periodic audio oscillation of high amplitude (known as the *Larsen effect*). In such a case, the electrical energy comes from the wall socket. If the amplifier is unplugged, there is no antidamping, and the oscillations will cease. Such hidden energy sources will characterize nearly all of the examples in this book.

1.5 Chaos and Strange Attractors

An important feature of flows in two-dimensional state space is that the streamlines cannot cross except at an X-point, which is always an unstable equilibrium with a vanishing flow. The implication of this fact is that the preceding figures represent all the types of bounded flows that can occur in two-dimensional systems of ordinary differential equations. In particular, chaos cannot occur (However, see Chapter 5 for some exceptions). Chaos requires at least three dimensions so that the streamlines can cross by passing behind one another. This notion has been formalized in the *Poincaré–Bendixson theorem* (Hirsch *et al.*, 2004).

As an example of such a three-dimensional flow, consider the simple damped pendulum but with a sinusoidal forcing given by

$$\dot{x} = v$$
$$\dot{v} = -bv - \sin x + \sin \Omega t,$$

(1.4)

where Ω is the frequency of the forcing, which is another parameter of the system. Such a system in which t appears explicitly on the right-hand side of the equations is said to be *nonautonomous*, but the time dependence can always be removed by adding another equation as in

$$\dot{x} = v$$
$$\dot{v} = -bv - \sin x + \sin z$$
$$\dot{z} = \Omega,$$

(1.5)

where z is the phase of the forcing function, which can be restricted to the range $0 \leq z < 2\pi$ (or $-\pi \leq z < \pi$) as was the case with x. The state space is now three-dimensional, and the governing equations are *autonomous* since t does not appear explicitly. Autonomous systems are important and desirable because there is a unique direction of the flow at almost every point in their state space independent of when the point is visited.

Note that the $\sin \Omega t$ in Eq. (1.4) could equally well have been replaced by $\cos \Omega t$ since the difference just amounts to starting the clock at a different time. Indeed, an arbitrary phase angle ϕ could have been added to give $\sin(\Omega t + \phi)$ without changing the dynamics. In terms of Eq. (1.5), the phase can be accounted for by changing the initial condition z_0. Furthermore, Ω can be restricted to positive values since the sine function has the same shape whether its argument is positive or negative.

Figure 1.6 shows the streamlines for such a flow with $b = 0.05$ and $\Omega = 0.8$ for an initial condition of $(x_0, v_0, z_0) = (0, 1, 0)$ projected onto the xv-plane with the z variable shown in shades of gray and with a subtle shadow to give the illusion of depth. The complicated and tangled behavior is a signature of chaos, and the trajectory winds around forever, never repeating, on an object called a *strange attractor*. Such an attractor has a noninteger dimension greater than 2.0 but less than 3.0. Only a portion of the state space trajectory is shown since it would otherwise densely cover a region of the plane and the individual streamlines would not be visible. As with the limit cycle, almost all initial conditions are drawn to the strange attractor. Dissipative chaotic flows generally have such strange attractors. Said differently, the appearance of a strange attractor in the state space dynamics is a signature of chaos, and they have been called the 'fingerprints of chaos' (Richards, 1999).

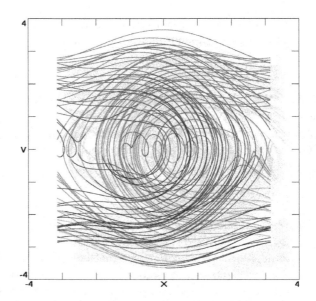

Fig. 1.6   State space plot showing a strange attractor for the damped forced pendulum in Eq. (1.5) with $b = 0.05$ and $\Omega = 0.8$ for an initial condition of $(x_0, v_0, z_0) = (0, 1, 0)$, $\lambda = (0.1400, 0, -0.1900)$.

Strange attractors are also called *chaotic attractors*, although the two terms are not precisely the same. There are examples of strange attractors that are not chaotic (Grebogi *et al.*, 1984; Ditto *et al.*, 1990; Feudel *et al.*, 2006) and chaotic attractors that are not strange (Anishchenko and Strelkova, 1997), but they are rare. The terminology depends on whether the geometric (strange) or dynamic (chaotic) property of the attractor is of primary interest. All the attractors in the remainder of this book are both strange and chaotic.

1.6   Poincaré Sections and Fractals

Given that the strange attractor for a three-dimensional chaotic flow must have a dimension greater than 2.0, it is useful to devise a method for displaying it in a two-dimensional plot. The *Poincaré section* provides such a method. Instead of projecting the entire state space trajectory onto the xv-plane as was done in Fig. 1.6, one can sample it at a particular value of the third variable (z in this case). Imagine that the three-dimensional trajectory in state space is illuminated briefly and periodically

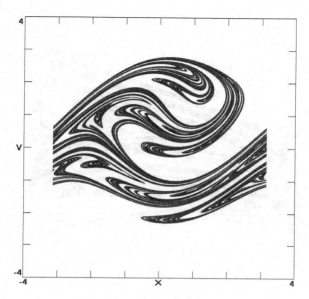

Fig. 1.7   Poincaré section for the damped forced pendulum in Eq. (1.5) with $b = 0.05$ and $\Omega = 0.8$ at z (mod 2π) $= \pi/2$.

with a strobe light synchronized with the forcing function. This method amounts to taking a plane slice through the strange attractor at a fixed value of the phase z, which reduces the dimension of the attractor by 1.0. Since the attractor for a dissipative three-dimensional chaotic flow must lie between 2 and 3, its Poincaré section will have a dimension between 1 and 2. Figure 1.7 shows such a Poincaré section for Eq. (1.5) with $b = 0.05$ and $\Omega = 0.8$ at the times when z (mod 2π) $= \pi/2$. Other values of z give similar results. In fact, a succession of Poincaré sections at successively increasing values of z can be used as frames of an endlessly repeating animation, some examples of which can be found at http://sprott.physics.wisc.edu/fractals/animated/.

A practical issue when calculating a Poincaré section is that the chosen time does not usually occur exactly at one of the finite iteration steps, but rather between two of them. In such a case, it is generally necessary and usually suffices to interpolate linearly between the two adjacent time steps. Failing to do so will smear out the fine-scale detail and will usually result in ugly plots. Of course it also helps to use relatively small time steps. Other more accurate methods are also available such as one suggested by Hénon (1982).

Fig. 1.8   Poincaré section for the damped forced pendulum in Eq. (1.5) with $b = 0.05$ and $\Omega = 0.8$ at $z \pmod{2\pi} = \pi/2$ as in Fig. 1.7 but wrapped onto a cylinder.

A Poincaré section can also be wrapped onto a cylinder as was done with the streamlines in Fig. 1.3. Figure 1.8 shows the result for the Poincaré section in Fig. 1.7. Note that there are many other sections that could have been taken, for example by fixing the angle of the pendulum (x) and plotting the velocity of the pendulum and phase of the forcing function on the surface of the cylinder, with a similar result.

Figures 1.7 and 1.8 illustrate why the attractor is *strange* and what it means to have a noninteger dimension. This object is an example of a *fractal*, about which books have been written, for example Mandelbrot (1983). This particular fractal has a dimension of about 1.7, which implies that the strange attractor for which it is a section has a dimension of about 2.7. The Poincaré section is somehow more than a long curvy line, but less than a surface with holes in it. Since its dimension is less than 2.0, it has no area. If one were to throw infinitely sharp darts at it, they would all miss. By the same token, it could not be seen except for the fact that the infinite collection of points that make it up are drawn with a nonzero size due to the limitation of computer graphics in the same way that a true mathematical line with zero width would not be visible if drawn on

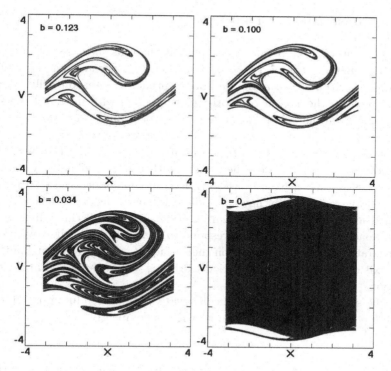

Fig. 1.9   Poincaré sections for the damped forced pendulum in Eq. (1.5) with $\Omega = 0.8$ for various values of damping (b) at z (mod 2π) $= \pi/2$ showing how the dimension increases from 'line-like' to 'surface-like' with decreased damping.

a surface even though it contains infinitely many points. The dimension of the Poincaré section is the same no matter what section is taken with rare exceptions, and wrapping the object onto a cylinder as in Fig. 1.8 also preserves the dimension.

It is often the case that the dimension of a strange attractor increases as the damping is reduced, approaching an integer value in the limit of zero damping, and such is the case for this example as Fig. 1.9 shows. The four Poincaré sections have approximate dimensions of 1.4, 1.6, 1.8, and 2.0, respectively, and represent a transition from an object that is 'line-like' (dimension 1.0) to 'surface-like' (dimension 2.0). However, it is rare for the dimension of an attractor to increase continuously and monotonically as the damping is reduced while remaining chaotic all the way to the limit of zero damping.

1.7   Conservative Chaos

Systems without damping can also exhibit chaos as the case with $b = 0$ in
Fig. 1.9 shows, but they do not have attractors. Instead, some initial conditions lie within the *chaotic sea* and eventually come arbitrarily close to
every point in the sea. The chaotic sea is a region with integer dimension,
although it can have infinitely many holes (or islands) of various sizes. If
the number of islands of each size scales as a power of their size, like craters
on the Moon, the chaotic sea is an example of a *fat fractal* (Farmer, 1985;
Grebogi *et al.*, 1985), also called 'dusts with positive volume' by Mandelbrot (1983). Initial conditions within these islands or otherwise outside
the chaotic sea typically have periodic orbits as indicated by closed loops
(called *drift rings*) in the Poincaré section. In the full three-dimensional
state space, these drift rings correspond to quasiperiodic orbits (two incommensurate frequencies) that lie on the surface of a 2-torus (a doughnut), on
the axis of which is a simple periodic orbit as will be described in the next
section.

The simplest example of the conservative forced pendulum is Eq. (1.4)
with $b = 0$ and $\Omega = 1$, which can be written compactly as

$$\ddot{x} + \sin x = \sin t, \qquad (1.6)$$

where $\ddot{x} = d^2x/dt^2 = d\dot{x}/dt = dv/dt$. Its Poincaré section at z (mod
$2\pi) = \pi/2$ is shown in Fig. 1.10 for various initial conditions. Remember
that a chaotic sea whose Poincaré section is two-dimensional is actually a
three-dimensional region of chaos sampled in cross section. That a system
as simple and elegant as Eq. (1.6) can have a chaotic solution as complicated
as Fig. 1.10 is the theme of this book.

Note that when a system is said to be *conservative*, it does not necessarily mean that the mechanical energy is conserved, but only that there
is no damping (friction). In fact the mechanical energy of the forced pendulum described by Eq. (1.6) varies drastically in time as Fig. 1.11 shows.
When averaged over a sufficiently long time, the energy is conserved, but
it can vary from zero to a rather large value as the forcing function supplies and removes energy from the system. The forcing function does work
$W = x \sin t$ on the system for this case, and that work can be positive or
negative, but it averages to zero and causes the energy to fluctuate. If it
were an electrical circuit instead of a pendulum, we would say that the
circuit is purely *reactive*, storing energy temporarily but not dissipating
it through electrical resistance. Furthermore, in a dissipative system the

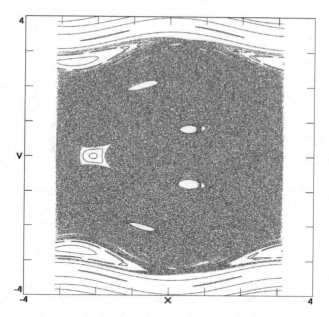

Fig. 1.10   Poincaré section at $z \pmod{2\pi} = \pi/2$ for the conservative forced pendulum in Eq. (1.6) for various initial conditions showing the chaotic sea with islands of periodicity.

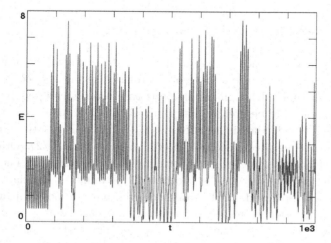

Fig. 1.11   Mechanical energy $E = v^2/2 + 1 - \cos x$ versus time for the conservative forced pendulum in Eq. (1.6) with initial conditions $(x_0, v_0, z_0) = (0, 1, 0)$ in the chaotic sea.

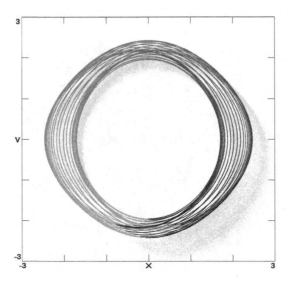

Fig. 1.12   State space plot showing an invariant 2-torus for the conservative forced pendulum in Eq. (1.6) with initial conditions $(x_0, v_0, z_0) = (-1.8, 0, \pi/2)$, $\lambda = (0, 0, 0)$.

energy decays to zero only when there is a point attractor. Otherwise, the forcing function or antidamping term supplies energy to replace that lost through dissipation, and the energy fluctuates about some average value as the trajectory moves around on the attractor.

1.8   Two-toruses and Quasiperiodicity

The holes in the Poincaré section in Fig. 1.10 represent regions in which the trajectories lie on the surface of a 2-torus, like the surface of a doughnut or inner tube, although the shape is only topologically equivalent to a doughnut and may be greatly distorted. Figure 1.12 shows the trajectory in the small hole in the vicinity of $(x, v) = (-2, 0)$ of Fig. 1.10. This trajectory is for an initial condition of $(x_0, v_0, z_0) = (-1.8, 0, \pi/2)$. Other initial conditions would lie on 2-toruses nested with the one shown. These nested toruses are successively smaller, ultimately forming an invariant circle at the center of the island. Each trajectory eventually comes arbitrarily close to every point on the surface of its torus, mapping out a two-dimensional surface in the three-dimensional state space. The surface has a cut where $z \pmod{2\pi} = 0$, but this cut could be removed by plotting the torus in cylindrical coordinates as was done in Fig. 1.3.

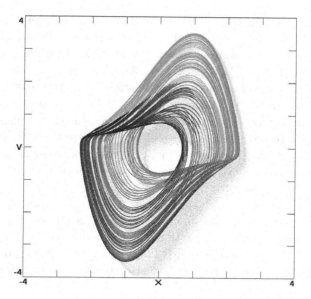

Fig. 1.13 State space plot showing an attracting 2-torus for the forced van der Pol oscillator in Eq. (1.7) with $A = 0.9$, $\Omega = 0.5$ and initial conditions $(x_0, v_0, z_0) = (1, 0.1, 0)$, $\lambda = (0, 0, -0.8264)$.

A 2-torus can also be attracting, just as a limit cycle, which can be considered as a 1-torus, is an attracting analog of the invariant circle. A simple example of an attracting 2-torus is the forced van der Pol oscillator

$$\ddot{x} + (x^2 - 1)\dot{x} + x = A \sin \Omega t \qquad (1.7)$$

with $A = 0.9$ and $\Omega = 0.5$. Its state space attractor as shown in Fig. 1.13 resembles a strange attractor, but it is not. Rather, the trajectory will eventually fill the entire two-dimensional surface of a torus, and hence it is not a fractal. For other values of the parameters, the trajectory closes on itself after a finite number of transits around the torus, and thus fails to fill the torus, producing a limit cycle. For yet other values of the parameters, the attracting torus develops wrinkles that eventually give way to a strange attractor.

For both the invariant torus and the attracting torus, the trajectory on the torus consists of two independent periodic motions, one the short way around the torus, and the other the long way. For the trajectory to fill the torus, the periods of these motions must be *incommensurate*, which means that their ratio is an irrational number (not a ratio of two integers). One would think that such a case would be overwhelmingly dominant, but for a nonlinear system like the forced van der Pol oscillator with two characteristic periods, the motion in xv-space tends to lock to a multiple of the period of the forcing function $(2\pi/\Omega)$, in which case there is only a single period producing a limit cycle (or an invariant circle in the case of a conservative system). When the periods are incommensurate, the motion is said to be *quasiperiodic*, and in the simplest case, it consists of a superposition of two sine waves of incommensurate frequencies. In higher dimensions, one can have 3-toruses, 4-toruses, and so forth, but they are less common than chaos.

As Figs. 1.12 and 1.13 suggest, it is often difficult to distinguish quasiperiodicity from chaos. In this case, taking a Poincaré section makes it easy, but when there are more than two frequencies or when the Poincaré section itself is not a simple closed loop, it can be difficult to distinguish between the two. Better methods are available as we now describe.

1.9 Largest Lyapunov Exponent

Chaos can often be identified with some confidence by observing the strange attractor or chaotic sea in a state space plot or Poincaré section. However, it is useful to have a more objective and quantitative measure of chaos, and for that purpose the *Lyapunov exponent* is ideal. The main defining feature of chaos is the sensitive dependence on initial conditions. Two nearby initial conditions on the attractor or in the chaotic sea separate by a distance that grows exponentially in time when averaged along the trajectory, leading to long-term unpredictability. The Lyapunov exponent is the average rate of growth of this distance, with a positive value signifying sensitive dependence (chaos), a zero value signifying periodicity (or quasiperiodicity), and a negative value signifying a stable equilibrium. The Lyapunov exponent is named in honor of the Russian mathematician Aleksandr Lyapunov who was one of the first to explore dynamic stability over a hundred years ago. His name is sometimes spelled as 'Liapunov.'

Calculation of the Lyapunov exponent is conceptually simple since one only needs to follow two initially nearby trajectories and fit the logarithm (base-e throughout this book) of their separation to a linear function of time. The slope of the fit is the Lyapunov exponent. However, there are a number of practical difficulties.

First, one has to follow the trajectory for a sufficiently long time to be sure it is on the attractor before beginning to calculate the exponent. There is no way to know for sure how long this will take, and there are situations, thankfully somewhat rare, where the time required is infinite. For a conservative system, it is not necessary to wait since there is no attractor, but one needs to ensure that the initial condition lies in the chaotic sea and is not on one of the equilibrium points or stable manifolds.

Second, the initial conditions chosen for the two trajectories will not in general be oriented in the direction of the most rapid expansion, and their distance may in fact initially decrease before beginning to increase. Fortunately, they tend to orient in the direction of the most rapid separation (or the least rapid contraction) rather quickly (but not always!). Therefore, it is advisable to discard some of the early points from the fit. There is no harm in letting this reorientation occur while the trajectory is approaching the attractor, thereby accomplishing two things at once.

Third is the fact that the rate of separation usually varies considerably with position on the attractor or in the chaotic sea, and thus it is necessary to average over a long time, to achieve a proper weighting of the points in each region of the attractor or sea. Since the calculated value usually converges slowly, it is useful to develop some criterion for convergence as will be described shortly.

Finally, and most seriously, is the fact that the initial conditions must have a separation large enough to be expressed accurately in whatever precision the computer is using, while not shrinking to zero or growing to a significant fraction of the size of the attractor or sea. This requirement is nearly always incompatible with the requirement of following the trajectories long enough for the calculated value to converge.

This final problem is solved by moving the trajectories back to some appropriate small separation along the direction of their separation whenever they get too far apart or too close together. If the separation is readjusted to d_0 after a time δt (which can also be the integration step size) and the separation at the end of the time step is d_1, the Lyapunov exponent is given by $\lambda = \langle \log(d_1/d_0) \rangle / \delta t$, where the angle brackets $\langle \rangle$ denote an average along the trajectory. The *ergodic hypothesis* (Ruelle, 1976) asserts that a time average along a single representative trajectory is equivalent to a spatial average over the attractor, weighted by the density of points on the attractor.

Figure 1.14 shows a graph of the Lyapunov exponent versus b for the damped, forced pendulum in Eq. (1.5) with $\Omega = 0.8$ and initial conditions

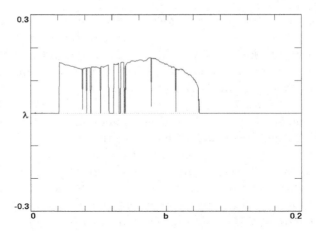

Fig. 1.14   Largest Lyapunov exponent for the damped, forced pendulum in Eq. (1.5) with $\Omega = 0.8$ and initial conditions $(x_0, v_0, z_0) = (0, 1, 0)$ as a function of the bifurcation parameter b.

$(x_0, v_0, z_0) = (0, 1, 0)$. It is common to see considerable structure in such plots since the Lyapunov exponent is usually not a smooth function of the parameters, with the discontinuities representing *bifurcations*, about which books have been written such as Wiggens (1988) and Kuznetsov (1995). However, it is easy to identify regions of positive and zero Lyapunov exponents and to confirm that the fractal Poincaré sections in Fig. 1.9 have positive Lyapunov exponents as expected for chaos, whereas the quasiperiodic trajectories such as the one in Fig. 1.12 have zero Lyapunov exponents (to within the precision of the calculation). Note the ubiquity of periodic windows in the chaotic region as is typical for low-dimensional chaotic systems. In fact, it is usually the case that there are infinitely many such windows of ever smaller widths, but fortunately there are also infinitely many values of the bifurcation parameter (b in this case) that lie outside these windows.

Graphs such as Fig. 1.14 are useful for studying the route to chaos and for identifying parameters that give chaos, but they are most useful for systems with a single adjustable parameter. When there are two or more parameters such as b and Ω in Eq. (1.5), it is more convenient and instructive to plot the regions in which the Lyapunov exponent is positive, zero, and negative in the plane of two of the parameters. Figure 1.15 shows such a plot for the damped, forced pendulum in Eq. (1.5), with black indicating chaos (C) and light gray indicating periodicity (P).

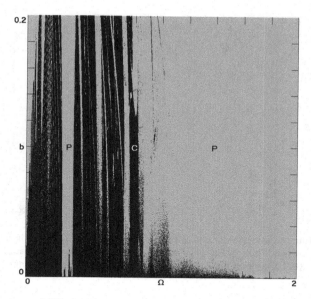

Fig. 1.15   Regions of different dynamics for the damped, forced pendulum in Eq. (1.5) with initial conditions $(x_0, v_0, z_0) = (0, 1, 0)$ as a function of the bifurcation parameters b and Ω. The chaotic regions (C) are shown in black, and the periodic regions (P) are shown in light gray.

A practical computational difficulty is distinguishing an exponent of zero from one that is small but nonzero. For the purpose of this book, values of the Lyapunov exponent for which $|\lambda| < 0.001$ will be taken as zero provided the estimated uncertainty in λ is less than $0.001 - |\lambda|$. Conversely, the Lyapunov exponent is taken as nonzero if $|\lambda| > 0.001$ provided the estimated uncertainty in λ is less than $|\lambda| - 0.001$. The sign of λ is assumed to have been determined correctly whenever one or the other of these conditions is satisfied, but there are always cases very near the bifurcation boundaries where the dynamics cannot be reliably classified.

What remains is to describe a method for estimating the best value of λ and its uncertainty $\delta\lambda$. For that purpose, a running average of $\log(d_1/d_0)$ is calculated at each time step (typically $\delta t = 0.1$) and the most recent m such values are stored in an array (typically $m = 1 \times 10^4$) as shown in Fig. 1.16. The array is divided into three equal groups ($m/3$ points each), and the average value of λ is calculated for the early (λ_e), mid (λ_m), and late (λ_l) third, as shown by the small circles in Fig. 1.16. The Lyapunov exponent is assumed to converge as $a + b/(t + C)$, and a curve of

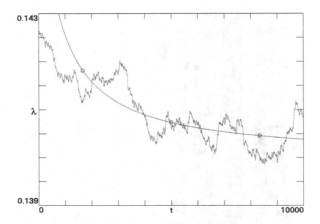

Fig. 1.16 Running average of the Lyapunov exponent over 1×10^4 time steps for the damped, forced pendulum in Eq. (1.5) with $b = 0.05$ and $\Omega = 0.8$ and initial conditions $(x_0, v_0, z_0) = (0, 1, 0)$, along with a curve fitted to the three average values shown with small circles.

that form is fitted to the data using least squares as shown by the smooth curve in Fig. 1.16, so that the extrapolated value for $t \to \infty$ is given by $\lambda = (2\lambda_e\lambda_l - \lambda_e\lambda_m - \lambda_m\lambda_l)/(\lambda_e - 2\lambda_m + \lambda_l)$ provided $|\lambda_l - \lambda_m| < |\lambda_m - \lambda_e|$. Otherwise, the Lyapunov exponent is not converging a and is estimated to be $\lambda = (\lambda_e + \lambda_m + \lambda_l)/3$. The uncertainty is assumed to be given by the range of values (largest minus smallest) stored in the array including the extrapolated value. This method has been tested and found to give reasonable results in cases where the exact Lyapunov exponent is known by other means, although the convergence is slower as the dimension of the attractor increases because it takes longer to sample the attractor or chaotic sea.

When λ is graphed versus a parameter as in Fig. 1.14, each data point is derived from a calculation that proceeds until $\delta\lambda$ is typically 10% the width of the line to ensure that the observed variations are real and not numerical. When values are quoted for the Lyapunov exponent, the uncertainty is typically less than 1×10^{-4} so that all four of the digits typically quoted are significant.

1.10 Lyapunov Exponent Spectrum

The previous discussion implied that there is a single Lyapunov exponent for a given flow. In fact, there are as many Lyapunov exponents as there

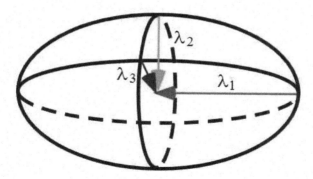

Fig. 1.17 A sphere of initial conditions distorts into an ellipsoid as time progresses, with each axis of the ellipsoid expanding at a rate given by one of the Lyapunov exponents (or contracting if that exponent is negative).

are state space variables, and what was calculated is only the largest (or least negative) of them. Fortunately, this is the only one that is required to identify chaos, since if it is positive, the system exhibits sensitive dependence on initial conditions independent of the values of the others, and if it is zero or negative, none of the others can be positive either.

The way to envision the Lyapunov exponents is to image a spherical ball filled with very many initial conditions in a three-dimensional state space. As the ball moves with the flow, it distorts into an ellipsoid as shown in Fig. 1.17, the longest axis of which expands at a rate given by the first Lyapunov exponent λ_1. Now consider a plane perpendicular to this expanding direction that cuts through the middle of the ellipsoid. In this plane, the points are contained within an ellipse whose major axis is expanding at a rate given by the second Lyapunov exponent λ_2 (or contracting if λ_2 is negative). The minor axis of this ellipse expands (or more likely contracts) at a rate given by the third Lyapunov exponent λ_3. From the definition, it is clear that $\lambda_1 \geq \lambda_2 \geq \lambda_3$, and this convention is universal.

As an example, Fig. 1.18 shows a succession of Poincaré sections for the conservative forced pendulum in Eq. (1.6) with a ball of radius $\sqrt{0.02}$ containing 10^6 initial conditions distributed throughout its volume. Even in a single transit ($N = 1$, corresponding to a time lapse of $\Delta t = 2\pi$), the ball has distorted into a highly elongated but very thin ellipsoid. In the direction perpendicular to the plane, there is neither expansion nor contraction since initial conditions distributed in that direction follow one another with a time lag that neither grows nor shrinks on average. Since the largest Lyapunov exponent is $\lambda_1 = 0.1491$, one would expect the ellipse

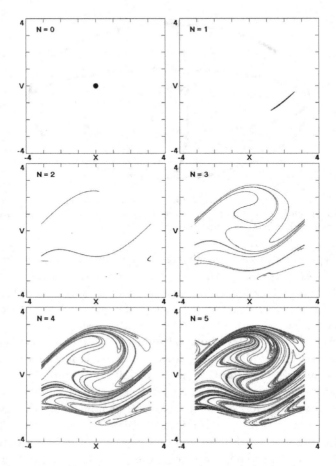

Fig. 1.18 A succession of Poincaré sections at z (mod 2π) $= \pi/2$ for the conservative forced pendulum in Eq. (1.6) with a ball of initial conditions, showing the filamentation of phase space that occurs as a result of the chaos.

to elongate by a factor of $e^{0.1491 \times 2\pi} \approx 2.55$ for each transit, and indeed it does on average, at least initially. However, the ellipse quickly distorts into a long, thin filament whose length grows much faster than expected because there are regions of state space near the separatrix where the local stretching is enormous. Consequently, after a very few transients, it has already begun to fill in much of the chaotic sea. The behavior is analogous to the way a drop of cream spreads throughout a cup of coffee after just a bit of stirring in a process called *filamentation*.

The idea generalizes to state spaces of any dimension, although it is hard to visualize dimensions greater than three. When we refer to *the* Lyapunov exponent λ, we will mean λ_1, although the Lyapunov exponent will usually be expressed as $\lambda = (\lambda_1, \lambda_2, \lambda_3, \ldots)$ as the occasion dictates. For example, a trajectory in the chaotic sea for Eq. (1.6) has $\lambda = (0.1491, 0, -0.1491)$. Note that the Lyapunov exponents for a conservative system are symmetric about zero because of the time-reversal invariance, and hence they necessarily sum to zero.

It is important to understand that as the ball moves with the flow, the orientation of the ellipsoid in state space is constantly changing and its axes are rarely aligned with the state space axes. Although there are as many Lyapunov exponents as there are state space variables, there is not a one-to-one correspondence between them. Furthermore, as the ellipsoid continues to stretch in one direction while contracting in another, it distorts into a long, thin filament as shown in Fig. 1.18 that eventually becomes a strange attractor as shown in cross section in Fig. 1.7 or a chaotic sea as shown in cross section in Fig. 1.10.

In a chaotic system, there must be stretching to cause the exponential separation of initial conditions but also folding to keep the trajectories from moving off to infinity. The folding requires that the equations of motion contain at least one nonlinearity, leading to the important principle that chaos is a property unique to nonlinear dynamical systems. If a system of equations has only linear terms, it cannot exhibit chaos no matter how complicated or high-dimensional it may be. However, nonlinearity does not guarantee chaos. For example, the $\sin \Omega t$ in Eq. (1.7) and elsewhere is a nonlinearity, but by itself cannot generate chaos. In fact, it can be replaced with two additional linear equations as will be shown in Chapter 6.

Note also that the volume of the ellipsoid is proportional to the product of its axes and thus increases according to $V = V_0 e^{(\lambda_1 + \lambda_2 + \lambda_3)t}$, so that $dV/dt = (\lambda_1 + \lambda_2 + \lambda_3)V$. The only way the volume can expand without limit is if the trajectory is unbounded and goes to infinity. The simplest example of such behavior is the one-dimensional flow given by the linear equation $\dot{x} = x$ whose solution is $x = x_0 e^t$. Two nearby initial conditions separate at the same exponential rate so that the one and only Lyapunov exponent is $\lambda = 1$. Even though the Lyapunov exponent is positive and there is sensitive dependence on initial conditions, we do not consider such cases to be chaotic. Chaotic systems must also be *recurrent*, which means that the trajectory will eventually return arbitrarily close to its starting

Table 1.1   Characteristics of the attractors for a bounded three-dimensional flow.

λ_1	λ_2	λ_3	Attractor	Dimension	Dynamic
−	−	−	Equilibrium point	0	Static
0	−	−	Limit cycle	1	Periodic
0	0	−	Attracting 2-torus	2	Quasiperiodic
0	0	0	Invariant torus	1 or 2	(Quasi)periodic
+	0	−	Strange	2 to 3	Chaotic

point (but not exactly since then it would then be periodic) and will do so repeatedly (Smale, 1967).

In practice, unbounded solutions in a model are detected and discarded whenever the absolute value of any of the state space variables exceeds some large value such as 1000. We are interested only in dissipative systems whose volume contracts onto an attractor or in conservative systems whose volume remains constant although perhaps stretching and thinning to such a degree that every point in the ball of initial conditions eventually comes arbitrarily close to any point in the chaotic sea.

Another property of bounded systems is that, unless the trajectory attracts to an equilibrium point where it stalls and remains forever, the points must continue moving forever with the flow. However, if we consider two initial conditions separated by a small distance along the direction of the flow, they will maintain their average separation forever since they are subject to the exact same flow but only delayed slightly in time. This fact implies that one of the Lyapunov exponents for a bounded continuous flow must be zero unless the flow attracts to a stable equilibrium.

The spectrum of Lyapunov exponents contains more information about the dynamics than does the largest exponent by itself. Considering that each exponent can be negative, zero, or positive, that their sum cannot be positive for a bounded system, and that at least one exponent must be zero except for a point attractor, there are five possible combinations for a three-dimensional state space as given in Table 1.1. In higher-dimensional state spaces, other cases are possible including attracting 3-toruses $(0, 0, 0, -, \ldots)$ and hyperchaos $(+, +, 0, -, \ldots)$.

What remains is the task of calculating the exponents other than λ_1. There are numerical procedures for doing this (Geist *et al.*, 1990), but they tend to be cumbersome and slow. Fortunately, there is a shortcut for three-dimensional chaotic flows. Since one exponent must be positive and one must be zero, the third λ_3 must be negative. Therefore, if λ_1 and the sum of the exponents are known, then λ_3 can be determined from $\lambda_3 = \sum \lambda - \lambda_1$.

As mentioned earlier, the sum of the exponents is the rate of volume expansion (or contraction when negative as is usual) and is given by $\Sigma\lambda = \langle \partial\dot{x}/\partial x + \partial\dot{y}/\partial y + \partial\dot{z}/\partial z \rangle$ averaged along the trajectory. For example, the case in Eq. (1.5) is simply given by $\Sigma\lambda = \partial\dot{v}/\partial v = -b$, emphasizing that dissipation is the reason for state space contraction. Thus knowing that $\lambda_1 = 0.1400$ in Fig. 1.6 leads immediately to $\lambda = (0.1400, 0, -0.1900)$ since $b = 0.05$.

For four-dimensional conservative chaotic systems, the Lyapunov exponents occur in equal and opposite pairs, which means that there must be two zero exponents $\lambda_2 = \lambda_3 = 0$, and the lone negative exponent is then equal to the negative of the positive one $\lambda_4 = -\lambda_1$. For four-dimensional dissipative systems and for all systems with dimension greater than four, these methods fail, and one must resort to a more difficult numerical calculation of the spectrum of exponents. In those few cases where that is necessary in this book, the method described by Wolf *et al.* (1985) has been used.

1.11 Attractor Dimension

The spectrum of Lyapunov exponents allows one to calculate the dimension of the attractor. It is reasonable that when all the exponents are negative, corresponding to contraction in all directions, then the attractor must be a point with a dimension of zero. Think of an ellipsoid all of whose axes are shrinking. When the largest exponent is zero and all the others are negative, the only noncontracting direction is parallel to the flow, and hence the attractor must be a limit cycle with a dimension of one. When the largest two exponents are zero and all the rest are negative, the trajectory can move freely in one direction perpendicular to the flow, but contracts in the other perpendicular directions, in which case the attractor is a 2-torus, a surface with a dimension of two. The number of leading zero exponents represents the number of directions that the trajectory can fill without contracting or expanding and hence is the dimension of the attractor.

If λ_1 is positive and λ_2 is zero, then there is a surface defined by these two perpendicular directions in which a cluster of initial conditions expands without limit. Thus the attractor can have a dimension no less than two. However, if the sum of the first three exponents is negative, the volume

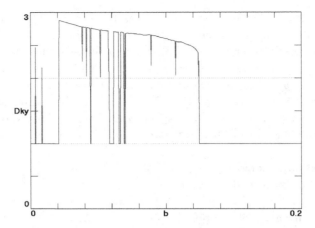

Fig. 1.19 Kaplan–Yorke dimension for the damped, forced pendulum in Eq. (1.5) with $\Omega = 0.8$ and initial conditions $(x_0, v_0, z_0) = (0, 1, 0)$ as a function of the damping b.

of a cluster of initial conditions must contract without limit, and so the attractor must have a dimension less than three. It is reasonable then to assign it a noninteger value between two and three in such a way that the cluster of initial conditions neither expands nor contracts in this fractional dimension.

The simplest way to do this is with a linear interpolation as suggested by Kaplan and Yorke (1979) and given by $D_{KY} = 2 - \lambda_1/\lambda_3$ in a three-dimensional state space. It is this argument that led to the estimates of the dimensions of the Poincaré sections in Fig. 1.9. As an example, Fig. 1.19 shows the Kaplan–Yorke dimension for the case in Eq. (1.5) with $\Omega = 0.8$ as a function of the damping b. The method easily generalizes to state spaces with dimensions greater than three. We will not generally calculate the Kaplan–Yorke dimensions for the cases presented in this book, but it is a simple matter to determine them with a pocket calculator from the spectrum of Lyapunov exponents, which is usually given.

Note that we now have three different dimensions to characterize a dynamical system. We have the number of degrees of freedom (1 for the pendulum), the dimension of the state space (2 for the free pendulum and 3 for the forced pendulum), and the dimension of the attractor (between 2 and 3 for the chaotic forced pendulum, depending on the damping). In addition, we could add the dimension of the parameter space (2 corresponding to b and Ω for the damped, forced pendulum).

1.12 Chaotic Transients

The alert reader may have noticed a slight discrepancy between Fig. 1.14 and Fig. 1.15 at small values of b where the former shows a periodic window with $\lambda = 0$ that is absent in the latter at $\Omega = 0.8$. The main difference between the two calculations is the time over which the trajectories are followed and the precision of the estimated Lyapunov exponent. A two-dimensional plot such as Fig. 1.15 requires much more computation (a factor of 800 in this case) than Fig. 1.14 for equivalent accuracy. Thus one would be inclined to accept the more accurate result in Fig. 1.14.

A careful examination of the case at $b = 0.01$ and $\Omega = 0.8$ where the two disagree shows that the discrepancy is caused by a chaotic transient. One way to verify and exhibit this effect is to make a plot similar to the Poincaré section in Fig. 1.7 but with v plotted versus the logarithm of time rather than versus x at those times when z (mod 2π) has a particular value such as $\pi/2$. This amounts to projecting the Poincaré section onto the v-axis so that a set of points projects onto another set of points. Any section with a dimension greater than or equal to one will project onto a line. Thus, if the attractor is a limit cycle with a dimension of one, its Poincaré section has a dimension of zero, whereas a chaotic system will have a dimension greater than two with a Poincaré section that projects onto a line.

Figure 1.20 shows such a plot. The logarithm of time is used so that a wide range of times can be displayed on the same plot. It is evident that the trajectory is chaotic until $t = 3.2 \times 10^4$, whereupon it abruptly becomes periodic and thereafter remains so. This is an example of *transient chaos*, and it is not particularly rare, although such very long transients typically occur only in small regions of parameter space. Edward Spiegel refers to transient chaos as 'pandemonium' (Smith, 2007).

The implication of such chaotic transients is that one has to be very cautious in declaring that a system has a chaotic attractor based solely on a numerical computation of its trajectory, even if the Lyapunov exponent appears to have converged to a large positive value. Most of the examples of chaos in this book have been verified by calculating for 2×10^9 iterations, corresponding typically to a time of $t = 2 \times 10^8$. The exceptions are the PDEs, DDEs, and other very high-dimensional systems, which are very computationally intensive. Another way to avoid being fooled by a chaotic transient is to see if the chaos persists over a range of parameters and initial conditions.

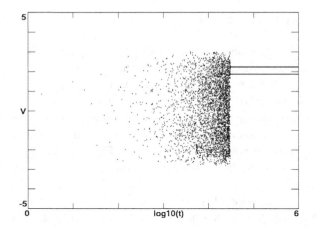

Fig. 1.20   Bifurcation diagram at $z \pmod{2\pi} = \pi/2$ versus time showing transient chaos for the damped, forced pendulum in Eq. (1.5) with $b = 0.01$ and $\Omega = 0.8$ for initial conditions $(x_0, v_0, z_0) = (0, 1, 0)$.

1.13   Intermittency

A phenomenon related to transient chaos is *intermittency* in which long periods of periodicity are interrupted by brief bursts of chaos, or perhaps the reverse (Pomeau and Manneville, 1980). Such systems are legitimately (though often only weakly) chaotic but require a long calculation for their Lyapunov exponent to converge. Intermittency is detected in the same way as transient chaos. Figure 1.21 shows an example of intermittency in the damped, forced pendulum with $b = 0.124304$ and $\Omega = 0.8$. The time scale here is linear rather than logarithmic. Fortunately, such behavior tends to occur only for very carefully chosen values of the parameters as is the case here, and so one is unlikely to stumble across it accidentally.

In searching for chaos, it is possible to have both false positives and false negatives, and one can never be completely confident that a system is chaotic based solely on numerical evidence, although considerable care has been taken to ensure that the chaotic examples in this book are not likely to be transients.

1.14   Basins of Attraction

The foregoing discussion assumed that a given dynamical system has only a single attractor for a given set of parameters, and thus by the ergodic

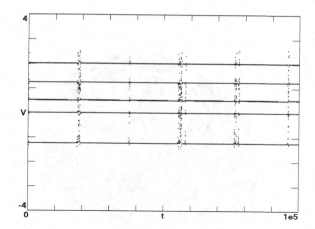

Fig. 1.21 Bifurcation diagram at $z \pmod{2\pi} = \pi/2$ versus time showing intermittency for the damped, forced pendulum in Eq. (1.5) with $b = 0.124304$ and $\Omega = 0.8$ for initial conditions $(x_0, v_0, z_0) = (0, 1, 0)$, $\lambda = (0.0037, 0, -0.1280)$.

hypothesis that the Lyapunov exponent is the same for all initial conditions. In fact, many dynamical systems have multiple coexisting attractors, sometimes infinitely many. Furthermore, even when a single attractor is present, some initial conditions may go to infinity rather than to the attractor, in which case infinity can be considered as an attractor. The various attractors can have quite different properties such as Lyapunov exponents and dimension, and can occur in all combinations — equilibrium points, limit cycles, toruses, and strange attractors. The number and type of attractors can change drastically as the parameters are varied.

Every attractor is contained within a *basin of attraction*, which is the region of state space over which initial conditions approach the attractor as $t \to \infty$. A useful analogy in two dimensions is a watershed representing those points on the Earth's surface where rainwater ends up in a given lake, or a wash basin that collects water from the faucet (tap in the UK) and directs it down the drain (or sink). Since chaos usually requires a three-dimensional state space, the basins for strange attractors are volumes (or hypervolumes in higher dimensions) inside of which the attractor resides. The basins can stretch to infinity, but more typically they have a boundary that may itself be a fractal (Aguirre *et al.*, 2009).

The damped, forced pendulum has limit cycles that coexist with the strange attractor for certain values of the parameters such as $0.0210 < b < 0.0275$ with $\Omega = 0.8$. Just as it is difficult to display an attractor

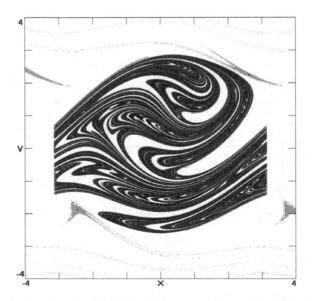

Fig. 1.22   Poincaré section for the damped, forced pendulum in Eq. (1.5) with $b = 0.024$
and $\Omega = 0.8$ at $z \pmod{2\pi} = \pi/2$ showing in gray the basin of attraction for the limit
cycles that coexist with the strange attractor and shown in cross section by the two small
black dots.

whose dimension is greater than 2.0 in a two-dimensional plot, it is even
more difficult to display its basin of attraction, whose dimension is three
or greater. However, one can plot a cross section of the basin in the same
way the Poincaré section was plotted in Fig. 1.7. In fact, the plots can be
overlaid since the variables are the same but evaluated at different times.
Figure 1.22 shows such a plot for the damped, forced pendulum in Eq. (1.5)
with $b = 0.024$ and $\Omega = 0.8$ at $z \pmod{2\pi} = \pi/2$. The small regions in gray
with long, thin tails represent those initial conditions that go to each of a
pair of limit cycles (represented by two small black dots, slightly enlarged
for enhanced clarity) rather than to the strange attractor whose Poincaré
section dominates the central part of the figure.

Figure 1.23 shows the two limit cycles superimposed on a Poincaré section of the strange attractor. The limit cycles correspond to unidirectional
rotations of the pendulum in opposite directions, and they each repeat after
three transits around the cylinder (a spatial periodicity in x of 6π).

Multiple coexisting attractors are more likely to occur when the damping is small, and when they do exist, it is likely that most if not all of

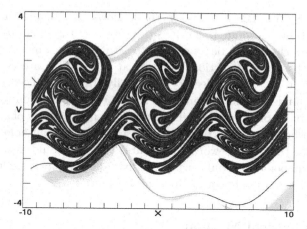

Fig. 1.23   Poincaré section for the damped, forced pendulum in Eq. (1.5) with $b = 0.024$ and $\Omega = 0.8$ at $z \pmod{2\pi} = \pi/2$ showing also projections of the two limit cycles onto the xv-plane.

them will be limit cycles (Feudel and Grebogi, 2003). Coexisting strange attractors are less common (Battelino *et al.*, 1988). There can be hundreds of coexisting attractors with fractal basin boundaries that come very close to almost every point in the state space (Feudel *et al.*, 1996). Systems with multiple attractors are said to be *multistable*.

The possible existence of multiple attractors means that it is necessary to search different initial conditions as well as different parameters when determining whether a given dynamical system is capable of exhibiting chaos. Furthermore, it means that plots such as Figs. 1.15 and 1.19 that show the regions of parameter space over which different types of dynamics occur are not generally unique, but rather depend on the chosen initial conditions. For this reason, initial conditions are given for most of the figures in this book even though they are usually not very critical.

In some cases, the parameters are slowly changed in the plots without reinitializing the variables with each change so as to remain in the chaotic or bounded region over the widest range of parameters. Such plots will often exhibit *hysteresis* in which case a second branch of the curve will be evident when the parameter is scanned in the opposite direction. Multiple attractors can often be identified in this way, but a more general method is just to try many different initial conditions and calculate for each case some value such as the Lyapunov exponent or more simply the mean value of a quantity such as $\langle v^2 \rangle$ that is not likely to be the same for the different attractors (Sprott, 2006). This method allows the search for coexisting attractors to be automated, and it is the way Fig. 1.22 was produced.

Transient chaos can be viewed as a situation in which an attractor touches its basin of attraction but only at places that are rarely visited by the trajectory. The trajectory is initially drawn to the attractor and wanders around on it for a long time before eventually coming to a place outside the basin of attraction, whereupon it escapes. Think of a fly buzzing around in a box for a long time before discovering a small hole in the wall that leads to the outside world. Of course the hole could also be a small patch of flypaper that would bring the fly to a permanent halt, just as a stable equilibrium might for a transiently chaotic trajectory.

1.15   Numerical Methods

One of the reasons so much attention has been paid to discrete-time systems is that they can be solved with great precision using digital computers. On the other hand, the systems in this book are governed by differential equations whose solutions advance forward in continuous time and are much more poorly approximated by the numerical methods that are used to solve them, all of which essentially consist of reducing them to some nearly equivalent discrete-time system, with a finite but small time step. Analytic solutions are available for linear systems of ODEs and for a few nonlinear ones, but essentially all chaotic systems require numerical solutions. Many books have been written about numerical methods for solving ODEs, for example, Gear (1971) and Shampine and Gordon (1975).

Nearly all the systems in this book were solved using a fourth-order Runge–Kutta algorithm with adaptive step size similar to that described by Press *et al.* (2007). The fourth-order Runge–Kutta method is given by

$$\mathbf{k}_1 = \mathbf{F}(\mathbf{x}_t)\delta t$$
$$\mathbf{k}_2 = \mathbf{F}(\mathbf{x}_t + \mathbf{k}_1/2)\delta t$$
$$\mathbf{k}_3 = \mathbf{F}(\mathbf{x}_t + \mathbf{k}_2/2)\delta t \tag{1.8}$$
$$\mathbf{k}_4 = \mathbf{F}(\mathbf{x}_t + \mathbf{k}_3)\delta t$$
$$\mathbf{x}_{t+\delta t} = \mathbf{x}_t + \mathbf{k}_1/6 + \mathbf{k}_2/3 + \mathbf{k}_3/3 + \mathbf{k}_4/6,$$

where $\mathbf{x}$ is a vector, each component of which is one of the dynamical variables: $\mathbf{x} = (x, y, z, \ldots)$, and $\mathbf{F}$ is a vector function whose components are $\mathbf{F} = (\dot{x}, \dot{y}, \dot{z}, \ldots)$. Note that time does not appear explicitly on the right-hand side of Eq. (1.8) because it can always be removed using the method illustrated in Eq. (1.5).

The maximum step size is typically taken as $\delta t = 0.1$, but for each step, the system is solved twice, once using the entire step and once using two half steps. If the two results differ by more than 1×10^{-8} in any of the variables, then the step is broken into as many as 1000 smaller time steps as required to achieve an accuracy of 1×10^{-8} over the whole step. Especially problematic cases have been checked by further reducing the step size or by putting a smaller error tolerance (typically 1×10^{-12}) on each step.

The maximum step size of $\delta t = 0.1$ is more than adequate for most cases because the natural period of oscillation is typically 1 radian per second when the parameters are of order unity, and thus there is the order of $2\pi/\delta t \approx 63$ iterations per cycle. For Eq. (1.1), the frequency of small-amplitude oscillations is exactly 1 radian per second, and the frequency of oscillation for the van der Pol oscillator is also very close to that value. Similarly, for the forced cases, the forcing frequency is 1 radian per second whenever $\Omega = 1$.

Even so, it is important to have independent checks on the reliability of the computations. One check is to monitor the energy versus time for a frictionless case such as Eq. (1.1) where it should be conserved. Another is to see that periodic solutions actually return to their starting points. Yet another is to see that the Lyapunov exponents sum to zero for those cases that have no dissipation. Such tests provide confidence that the chaotic cases presented here are not numerical artifacts resulting from a flawed integration method as can otherwise easily happen.

1.16   Elegance

Unlike the material in the previous sections, *elegance* is not a mathematical term, but rather, like beauty, is in the eye of the beholder. The concept is hard to quantify, but as Supreme Court Justice Potter Stewart said of pornography, 'I know it when I see it.' Certainly Eq. (1.6) is elegant, whereas a system such as

$$\ddot{x} - (0.28 - 0.32x^2 + 0.01\dot{x}^2)\dot{x} + 0.63\sin(-1.24x - 0.45) + 1.55x^3$$

$$= 3.42\sin 0.74t, \tag{1.9}$$

which is also chaotic and possesses many of the same properties, is much less elegant. Its state space plot is shown in Fig. 1.24.

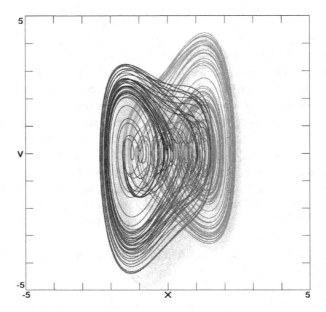

Fig. 1.24   State space plot showing a strange attractor for the very inelegant, damped, forced pendulum in Eq. (1.9) with an initial condition of $(x_0, v_0, z_0) = (1, 1, 0)$, $\lambda = (0.0978, 0, -0.1391)$.

While the term 'elegance' throughout this book refers to the form of the equation, many of the figures are elegant in their own right, but that is not what the book is about. The engaging plots are merely a delightful byproduct of the elegant equations, and there is no reason to believe that the elegance of the figure has any relation to the elegance of the equation that produced it. In fact, contrary to intuition, some of the most complicated dynamics arise from the simplest equations, while complicated equations often produce very simple and uninteresting dynamics. It is nearly impossible to look at a nonlinear equation and predict whether the solution will be chaotic or otherwise complicated. Small variations of a parameter can change a chaotic system into a periodic one, and vice versa.

A system of equations is deemed most elegant if it contains no unnecessary terms or parameters and if the parameters that remain have a minimum of digits. This notion can be quantified by writing an equation such as Eq. (1.9) in its most general form such as

$$\ddot{x} - (a_1 - a_2 x^2 - a_3 \dot{x}^2)\dot{x} + a_4 \sin(a_5 x + a_6) + a_7 x^3 = a_8 \sin a_9 t \quad (1.10)$$

and adjusting the parameters a_1 through a_9 to achieve this end. Ideally, we want as many of the parameters of be zero as possible while preserving the chaos, and the greatest number of those that remain should be ± 1. Note that it is generally possible to linearly rescale the variables (x and t in this case) so that a corresponding number of the parameters are ± 1.

One way to quantify the elegance of Eq. (1.10) is thus to count the number of nonzero parameters and then add to that count the total number of digits including the decimal point but excluding leading and trailing zeros for any parameters that are not ± 1 so that integer parameters are preferred. The resulting number, which perhaps should be called the *inelegance*, is the quantity to be minimized. By this criterion, Eq. (1.9) has an inelegance of 39, whereas Eq. (1.6), when viewed as a special case of Eq. (1.10) with $a_1 = a_2 = a_3 = a_6 = a_7 = 0$ and $a_4 = a_5 = a_8 = a_9 = 1$, has an inelegance of 4.

With such a quantitative measure of elegance, one can automate the optimization process, starting either from a known chaotic system or from one that is potentially chaotic. In this way, the method can be used to find new chaotic systems and then to simplify them. Unfortunately, the optimization is computationally intensive since there is little alternative to trial and error. In practice, the method starts from a known chaotic case or from an arbitrary place in parameter space and searches a random Gaussian neighborhood of the starting point, typically with a variance of 1.0, seeking a case with a Lyapunov exponent greater than 0.001. When a candidate is found, the search moves to that place in parameter space and continues searching from there. Each time a chaotic case is found, the search thereafter considers only cases whose elegance is at least as great as the best case found so far.

A system such as Eq. (1.10) typically converges to an optimum after a few thousand trials and about an hour of computation on a personal computer as Fig. 1.25 shows. One can achieve faster convergence at the expense of more false positives by relaxing the precision of the calculation. It also helps to reduce the number of parameters so that the search space has a lower dimension or by starting with an otherwise more elegant case. Sometimes it is useful to put constraints on the search space such as limiting the search to only positive values of the parameters or by imposing some special desired symmetry (Gilmore and Letellier, 2007). The cases in this book represent about ten CPU-years of searching and a similar amount verifying the results and calculating the Lyapunov exponents.

Of course, there may not be a single optimum since more than one

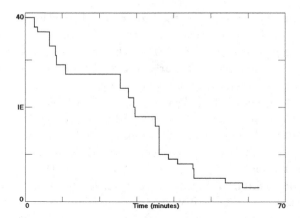

Fig. 1.25   Time history by the wall clock of the inelegance during a typical search for the maximally elegant form of Eq. (1.10) as given by Eq. (1.6) starting from the very inelegant form in Eq. (1.9) using a vintage 2008 personal computer.

combination of parameters may have the same elegance. In that event, the case with the largest Lyapunov exponent is usually chosen, or the case with coefficients closest to unity (2 is better than 3, and 0.9 is better than 0.8), or perhaps a case is chosen that has some nice symmetry such as equal values of two of the parameters, or simply one whose state space plot is more 'interesting.'

Just as one can find the most elegant set of parameters for a given system, it is possible to find the most elegant set of initial conditions within the basin of attraction or chaotic sea. However, it is usually more useful to have initial conditions that are close to the attractor to reduce the transients that would otherwise occur. Thus most of the initial conditions cited in this book are a compromise chosen to be close to the attractor (typically within a distance of 0.1) but using values of the initial conditions rounded to a few digits, while avoiding cases for which all the time derivatives are zero, which would imply an equilibrium point.

Finally, it is important to verify any chaotic system found in this way with a much longer, independent calculation of higher precision to avoid programming errors, transients, unboundedness, or other anomalies. Even so, there is no claim, much less guarantee, that the cases in this book are maximally elegant by the chosen criterion. Finding such cases is a goal, but failure to do so is not a serious flaw since the criterion is arbitrary. There is ample room for the interested reader to find cases yet more elegant than those presented here.

Chapter 2

Periodically Forced Systems

Some of the earliest and simplest examples of chaos occur in periodically forced nonlinear oscillators. An important example is the forced pendulum that was discussed in some detail in the previous chapter. This chapter will consider other periodically forced systems, some of which are even more elegant and potentially useful than the pendulum.

2.1 Van der Pol Oscillator

In the 1920s the Dutch physicist Balthasar van der Pol was working for Philips Glowlampworks, Ltd. in the Netherlands and became interested in relaxation oscillators, especially in the vacuum tube ('valve' in the UK) electronic circuits that had recently been developed. He proposed a mathematical model for such oscillators (van der Pol, 1920, 1926) now known as the van der Pol equation

$$\ddot{x} + b(x^2 - 1)\dot{x} + x = 0 \qquad (2.1)$$

and discussed in the previous chapter. This equation has been used to model heartbeats (van der Pol and van der Mark, 1928), pulsating stars called *Cephids* (Krogdahl, 1955), neuronal activity (Flaherty and Hoppensteadt, 1978), lobster chewing (Rowat and Selverston, 1993), earthquakes (Cartwright *et al.*, 1997), oscillations in a magnetized plasma column (Gyergyek *et al.*, 1997), vocal fold vibrations (Laje *et al.*, 2001), well drill chatter (Weinert *et al.*, 2002), bipolar disorders (Sprott, 2005), sunspot cycles (Passos and Lopes, 2008), and many other oscillatory phenomena.

Van der Pol was interested in the behavior of such oscillators when they are periodically forced according to

$$\ddot{x} + b(x^2 - 1)\dot{x} + x = A \sin \Omega t. \qquad (2.2)$$

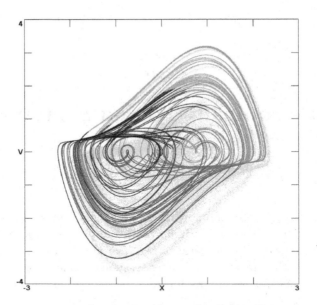

Fig. 2.1 Strange attractor for the forced van der Pol oscillator from Eq. (2.2) with $(b, A, \Omega) = (1, 1, 0.45)$ for initial conditions $(x_0, v_0, t_0) = (1, 1, 0)$, $\lambda = (0.0389, 0, -0.3262)$.

Van der Pol observed chaos in a periodically forced neon-bulb relaxation oscillator (see Chapter 10) for which this equation is a reasonable model.

Equation (2.2) was later studied carefully by Cartwright and Littlewood (1945) and was shown by Levinson (1949) to have what we now call chaotic solutions, long before the term 'chaos' was coined by Li and Yorke (1975). A more modern analysis of the system has been carried out by Levi (1981) and many others. The system typically exhibits limit cycles that are locked in frequency to submultiples of the forcing frequency Ω, but it can also produce 2-toruses, an example of which for $(b, A, \Omega) = (1, 0.9, 0.5)$ was shown in Fig. 1.13. Equation (2.2) exhibits chaos for $(b, A, \Omega) = (1, 1, 0.45)$, producing the strange space attractor shown in Fig. 2.1.

The term in parenthesis in Eq. (2.2) resembles $-\cos x$ since the expansion of the cosine for $|x| \ll 1$ is $\cos x = 1 - x^2/2 + \dots$. An elegant version of such a system is given by

$$\ddot{x} - 4\dot{x}\cos x + 2x = 5\sin 0.8t \tag{2.3}$$

with an attractor as shown in Fig. 2.2.

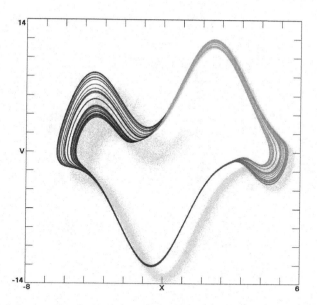

Fig. 2.2 Strange attractor for the modified forced van der Pol oscillator from Eq. (2.3) for initial conditions $(x_0, v_0, t_0) = (-0.9, 4, 0)$, $\lambda = (0.0836, 0, -0.9952)$.

2.2 Rayleigh Oscillator

The van der Pol equation (2.1) is closely related to the Rayleigh differential equation (Birkhoff and Rota, 1978) in which the x^2 term is replaced with $\dot{x}^2$. In fact the former can be derived from the latter by differentiating each term with respect to time and then replacing $\dot{x}$ with x. Each system has the property that the origin is an unstable equilibrium, and trajectories near it spiral outward until either x^2 or $\dot{x}^2$ gets sufficiently large, whereupon the antidamping is offset by the damping at large amplitude. An elegant version of the forced Rayleigh oscillator with chaotic solutions is

$$\ddot{x} + (\dot{x}^2 - 4)\dot{x} + x = 5\sin 4t \tag{2.4}$$

with an attractor as shown in Fig. 2.3.

2.3 Rayleigh Oscillator Variant

A much simpler variant of the Rayleigh Oscillator with two fewer terms is given by

$$\ddot{x} + x\dot{x}^2 = \sin 4t. \tag{2.5}$$

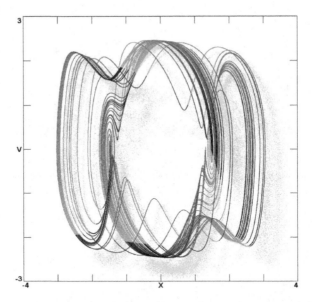

Fig. 2.3 Strange attractor for the forced Rayleigh oscillator from Eq. (2.4) for initial conditions $(x_0, v_0, t_0) = (1, 0.8, 0)$, $\lambda = (0.1542, 0, -3.9587)$.

This is a conservative system with the friction force proportional to the displacement x and to the square of the velocity $\dot{x}^2$. The energy lost when x is positive is cancelled by the energy gained when x is negative (antifriction) averaged over a cycle of the oscillation. Its state space plot is shown in Fig. 2.4.

2.4 Duffing Oscillator

In the previous examples, the nonlinearity responsible for the chaos was in the damping $(\dot{x})$ term. Even more common and more widely studied is the system in which the damping is linear but the restoring force contains a cubic nonlinearity as introduced by the German electrical engineer Georg Duffing (1918) and given most generally by

$$\ddot{x} + b\dot{x} + k_1 x + k_2 x^3 = A \sin \Omega t. \tag{2.6}$$

This system is a good model of various phenomena (Virgin, 2000) such as a magnetoelastic buckled beam (Moon and Holmes, 1979) or a nonlinear electronic circuit (Ueda, 1979). Most simply, it would model a mass on a spring that does not obey Hooke's Law, which states that the force exerted

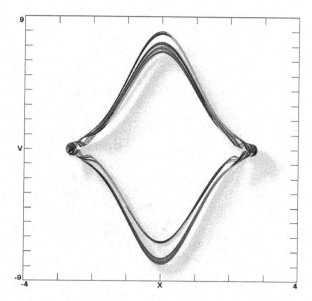

Fig. 2.4 State space plot for the Rayleigh oscillator variant from Eq. (2.5) for initial conditions $(x_0, v_0, t_0) = (0, 8, 0)$, $\lambda = (0.0132, 0, -0.0132)$.

by the spring is proportional to its stretch or compression but oppositely directed.

We can distinguish four cases for Eq. (2.6) depending on the signs of k_1 and k_2. If both signs are negative, then the solutions are unbounded and thus cannot lead to chaos. The case with $k_1 > 0$ and $k_2 > 0$ is called the *stiffening-spring* case because it models a spring that gets stiffer as it is stretched or compressed. The case with $k_1 > 0$ and $k_2 < 0$ is called the *softening-spring* case because it models a spring that gets weaker when it is stretched or compressed. It could also model a pendulum driven to large amplitude, but not so large that it goes 'over the top' since the expansion for the sine is $\sin x = x - x^3/6 + \ldots$ for $|x|$ small. The case with $k_1 < 0$ and $k_2 > 0$ is called *Duffing's two-well oscillator* because it would model a ball rolling in a trough with two dips separated by a bump. All three cases can exhibit chaos for appropriately chosen parameters, although the two-well case is the most common since its solutions are always bounded by the strong cubic restoring force.

An elegant example of chaos occurs in the damped two-well oscillator given by

$$\ddot{x} + \dot{x} - x + x^3 = \sin 0.8t \qquad (2.7)$$

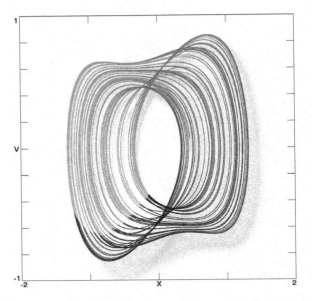

Fig. 2.5   Strange attractor for the damped Duffing two-well oscillator from Eq. (2.7) for initial conditions $(x_0, v_0, t_0) = (-1, -0.6, 0)$, $\lambda = (0.1216, 0, -1.1216)$.

whose attractor is shown in Fig. 2.5. A variant of this equation without the linear term in x was studied by Bonatto *et al.* (2008).

If we allow the damping to be zero ($b = 0$) in Eq. (2.6), then there are two even more elegant forms given by

$$\ddot{x} - x + x^3 = \sin t \tag{2.8}$$

and

$$\ddot{x} + x^3 = \sin 2t. \tag{2.9}$$

Equation (2.8) is a conservative two-well forced oscillator, while Eq. (2.9) is a single-well oscillator that has been studied by Gottlieb and Sprott (2001) who claim that it is the simplest forced chaotic oscillator, although it would have to vie against Eq. (1.6) for that distinction. The state space plots for these two systems are shown in Fig. 2.6.

It is possible to construct chaotic forced oscillators that are combinations of the cases described thus far. For example, a pendulum with van der Pol damping was considered in Eq. (1.3). The Rayleigh–Duffing oscillator was treated by Hayashi *et al.* (1970), and the Duffing-van der Pol oscillator and other combinations were studied by Yoshisuke Ueda (2001) using analog and digital computers as early as 1961 while he was a graduate student,

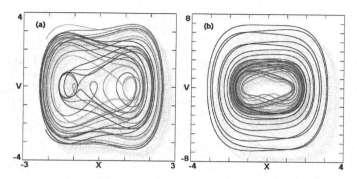

Fig. 2.6   Chaotic trajectories for the conservative forced Duffing oscillator (a) from Eq. (2.8) for initial conditions $(x_0, v_0, z_0) = (2, 1, 0)$, $\lambda = (0.1154, 0, -1.1154)$ and (b) from Eq. (2.9) for initial conditions $(x_0, v_0, t_0) = (1, 1, 0)$, $\lambda = (0.0907, 0, -0.0907)$.

but he was discouraged by his professor from publishing his observations of chaos for many years (Abraham and Ueda, 2000). The Rayleigh and van der Pol equations can be combined to give a limit cycle that is a perfect circle in state space (Sprott, 2003). However, these hybrid cases are less elegant than their component forms and tend not to be any more interesting. Indeed the case in Eq. (1.9) that combines all four nonlinearities and might be called a 'forced van der Pol–Rayleigh–Duffing pendulum' was used as an extreme example of inelegance.

2.5   Quadratic Oscillators

All of the examples considered so far have had been of the form

$$\ddot{x} + f(\dot{x}, x) = A \sin \Omega t \qquad (2.10)$$

where $f(\dot{x}, x)$ contains at least one sinusoidal or cubic nonlinearity, as dictated by a desire for the oscillator to have symmetry about its equilibrium. An obvious simplification is to ask whether *quadratic* functions $f(\dot{x}, x)$ can suffice to give chaos in such a forced oscillator. Four equally elegant cases are given in Table 2.1 with their corresponding attractors shown in Fig. 2.7.

Other systems with the form of Eq. (2.10) in which $f(\dot{x}, x)$ is a more complicated nonlinear function have been studied by Scheffczyk *et al.* (1991), but they are less elegant than the systems presented here. Systems in which $f(\dot{x}, x) = g(x)\dot{x} + h(x)$ are called forced *Liénard systems* and have a number special important properties, especially when $g(x)$ is an

Table 2.1   Chaotic forced damped quadratic oscillators.

Model	Equation	x_0, v_0, t_0	Lyapunov Exponents
FQ_1	$\ddot{x} + (1 - \dot{x})\dot{x} + 0.4x = \sin t$	$0, -0.4, 0$	$0.0493, 0, -1.0493$
FQ_2	$\ddot{x} + (x - \dot{x})\dot{x} + 0.3 = \sin t$	$1, -0.4, 0$	$0.0684, 0, -1.0609$
FQ_3	$\ddot{x} + (\dot{x} - x)\dot{x} + x = 0.3\sin t$	$0, 0.1, 0$	$0.0620, 0, -0.6106$
FQ_4	$\ddot{x} + (1 + 3x)\dot{x} + (1 - x)x = \sin t$	$0, -0.3, 0$	$0.0387, 0, -2.0604$

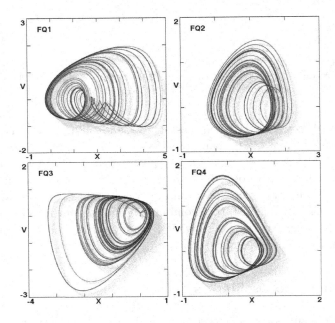

Fig. 2.7   Strange attractors for the forced damped quadratic oscillators in Table 2.1.

even function of x and $h(x)$ is an odd function of x as in the van der Pol and Duffing oscillators (Strogatz, 1994).

2.6   Piecewise-linear Oscillators

The existence of chaos in forced oscillators with quadratic nonlinearities suggests that it might also occur in piecewise-linear systems involving a single absolute value. Two rather simple cases are

$$\ddot{x} + |\dot{x} - x| + x - 1 = \sin t \qquad (2.11)$$

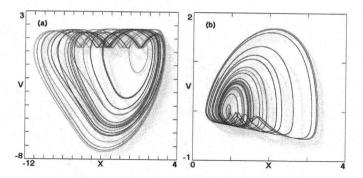

Fig. 2.8   Strange attractors for the piecewise-linear oscillators (a) from Eq. (2.11) for initial conditions $(x_0, v_0, z_0) = (0, 0.5, 0)$, $\lambda = (0.0812, 0, -0.6291)$ and (b) from Eq. (2.12) for initial conditions $(x_0, v_0, t_0) = (1, -0.2, 0)$, $\lambda = (0.0515, 0, -4.7749)$.

and

$$\ddot{x} + 3\dot{x} - 4|\dot{x}| + x = \sin t \qquad (2.12)$$

with attractors as shown in Fig. 2.8. Piecewise-linear oscillators with more than two linear regions that produce chaos when periodically forced have been extensively studied, for example using Chua's diode (Murali *et al.*, 1994), but for the present purpose they are deemed less elegant than systems that involve functions with only two linear regions and can thus be represented by a single absolute value nonlinearity.

2.7   Signum Oscillators

Whereas the absolute value is a *continuous* piecewise-linear function, discontinuous functions sometimes arise in models of real applications. An example is a harmonic oscillator with Coulomb damping, as would result from a friction force that is independent of velocity and position, like sliding friction and unlike air resistance, which increases with velocity (Den Hartog, 1930; Squire, 1985; Peters and Pritchett, 1997). The simplest such model for a forced oscillator would be Eq. (2.10) with $f(\dot{x}, x) = b \operatorname{sgn} \dot{x} + x$ where sgn is the signum function, $\operatorname{sgn} \dot{x} = \dot{x}/|\dot{x}|$, which is either $+1$ or -1 depending on whether its argument $\dot{x}$ is positive or negative, respectively. To be complete, we should add that $\operatorname{sgn} 0 = 0$. This system does not appear to admit chaotic solutions, nor does it exhibit chaos when the signum function is replaced with a similar but more gentle sigmoid function such as the hyperbolic tangent. The real situation is more complicated since several

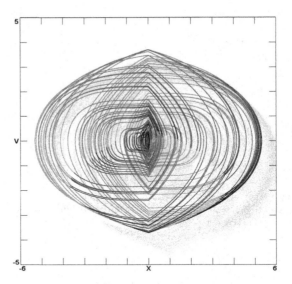

Fig. 2.9   State space plot for the signum system in Eq. (2.13) with initial conditions $(x_0, v_0, t_0) = (1, 0.1, 0)$, $\lambda = (0.0465, 0, -0.0465)$.

forms of friction usually coexist, and even the Coulomb friction takes on different values depending on position and on whether or not the velocity is zero, and chaos is common in such cases (Feeny, 1992; Feeny and Moon, 1994).

However, we earlier showed that if the x in $f(\dot{x}, x)$ is replaced by $\sin x$, the system is chaotic for $(b, A, \Omega) = (0, 1, 1)$ as indicated by Eq. (1.6), and if x is replaced by x^3, it is chaotic for $(b, A, \Omega) = (0, 1, 2)$ as indicated by Eq. (2.9). In such cases the chaos persists when a damping term $b\,\mathrm{sgn}\,\dot{x}$ is added, but here we are interested in cases where the only nonlinearity involves the signum function.

If the signum function is allowed to operate on x instead of $\dot{x}$, then chaotic solutions are possible, a particularly simple conservative example of which is

$$\ddot{x} + \mathrm{sgn}\,x = \sin t \tag{2.13}$$

whose state space plot is shown in Fig. 2.9. This system has an elegance comparable to Eq. (1.6) and Eq. (2.9).

For computational reasons (Grantham and Lee, 1993), the signum function is replaced with $\tanh(500x)$ for calculating the Lyapunov exponents (Gans, 1995). The value of 500 is not critical provided it is large but not

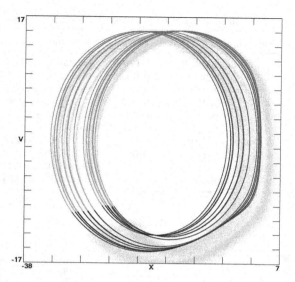

Fig. 2.10 Attractor for the forced exponential oscillator in Eq. (2.14) with $A = 20$ for initial conditions $(x_0, v_0, t_0) = (-24, -9, 0)$, $\lambda = (0.0181, 0, -1.0181)$.

so large that the iteration step carries the trajectory past the region where $|\tanh(500x)|$ is substantially less than 1.0 in a single time step.

2.8 Exponential Oscillators

Many other forced nonlinear oscillators are capable of producing chaos. A simple such system that frequently arises in electronic circuits that involve a diode is given by (van Buskirk and Jeffries, 1985)

$$\ddot{x} + \dot{x} + e^x - 1 = A \sin t, \tag{2.14}$$

which is chaotic for $|A|$ larger than about 19.7. Its attractor for $A = 20$ as shown in Fig. 2.10 can be considered as a limit cycle that drifts to the right until it bumps up against the strong nonlinearity that occurs for $x > 0$, whereupon it is chaotically kicked back to the left.

2.9 Other Undamped Oscillators

The previously discussed examples of Eq. (1.6), (2.8), and (2.9) suggest a family of systems of the form of Eq. (2.10) in which $f(\dot{x}, x)$ is a nonlinear function of x alone. Such systems would necessarily lack damping and thus

Table 2.2   Chaotic forced conservative nonlinear oscillators.

Model	Equation	x_0, v_0, t_0	Lyapunov Exponents
FC_0	$\ddot{x} + x^3 = \sin 2t$	$0, 0, 0$	$0.0920, 0, -0.0920$
FC_1	$\ddot{x} + x^5 = \sin 2t$	$0, 0, 0$	$0.1549, 0, -0.1549$
FC_2	$\ddot{x} + x^7 = \sin 2t$	$0, 0, 0$	$0.1895, 0, -0.1895$
FC_3	$\ddot{x} + x^9 = \sin 2t$	$0, 0, 0$	$0.1988, 0, -0.1988$
FC_4	$\ddot{x} + x^{11} = \sin 2t$	$0, 0, 0$	$0.2170, 0, -0.2170$
FC_5	$\ddot{x} + x/\sqrt{|x|} = \sin 2t$	$0.7, 0, 0$	$0.0352, 0, -0.0352$
FC_6	$\ddot{x} + x|x| = \sin 2t$	$0, 0.3, 0$	$0.0214, 0, -0.0214$
FC_7	$\ddot{x} + x|x|^3 = \sin t$	$0, 0, 0$	$0.0431, 0, -0.0431$
FC_8	$\ddot{x} + 0.2\sinh x = \sin t$	$1, 0, 0$	$0.0264, 0, -0.0264$
FC_9	$\ddot{x} + x - \tanh x = \sin t$	$1, 0, 0$	$0.0125, 0, -0.0125$

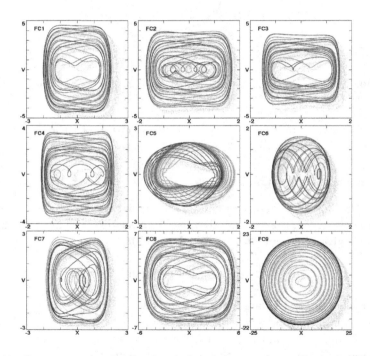

Fig. 2.11   State space plots for the chaotic forced conservative oscillators in Table 2.2.

would be conservative. Several such systems are listed in Table 2.2 (Sprott, 2003), with state space plots as shown in Fig. 2.11. The model FC_0 is the same as Eq. (2.9) and has already been discussed and illustrated in Fig. 2.6 but is included in the table for completeness.

Table 2.3   Chaotic velocity forced oscillators from Eq. (2.15) with $A = 1$ and $\Omega = 1$.

Model	Equation	x_0, v_0, t_0	Lyapunov Exponents
VF_1	$\dot{v} = -\sin x$	$0, 1, 0$	$0.1494, 0, -0.1494$
VF_2	$\dot{v} = (1 - x^2)v - 0.2x$	$0, 2, 0$	$0.1039, 0, -2.9872$
VF_3	$\dot{v} = v\cos 0.9x - 0.1x$	$2.7, 0, 0$	$0.0330, 0, -0.3357$
VF_4	$\dot{v} = x - x^3$	$0, 1, 0$	$0.1155, 0, -0.1155$
VF_5	$\dot{v} = -v + (1 + 0.6x)x$	$-2, 0, 0$	$0.0619, 0, -1.0619$
VF_6	$\dot{v} = -(1 + v)v - (1 - 2v)x$	$-3, 0.4, 0$	$0.0544, 0, -3.1682$
VF_7	$\dot{v} = -|v - 0.9x| - x$	$-2, 0.1, 0$	$0.0601, 0, -0.7578$
VF_8	$\dot{v} = -0.2v + \text{sgn}(v + x) - x$	$1, 0.7, 0$	$0.0088, 0, -0.1035$

2.10   Velocity Forced Oscillators

In all the previous examples, the periodic forcing acted on the $\dot{v}$ term as would normally be the case for a mechanical system in which the force obeys Newton's second law, $F = m\dot{v}$. However, that is not the only possibility. Shaw (1981) observed that the region of parameter space over which chaos occurs in the forced van der Pol oscillator is larger if the forcing acts instead on the $\dot{x}$ term. We will refer to this case as *velocity forcing* to distinguish it from the normal acceleration forcing of a mechanical system. Although such a condition would be unusual in a mechanical system, it can be readily achieved in electrical and other systems.

Velocity forcing is described by the system of equations

$$\dot{x} = v + A\sin\Omega t$$
$$\dot{v} = f(v, x) \tag{2.15}$$

where $f(v, x)$ could be any of the cases previously considered for acceleration forcing. Some examples of such systems are given in Table 2.3 with their corresponding state space plots in Fig. 2.12. Curiously, the velocity forced Rayleigh oscillator does not appear to admit chaotic solutions and thus is not represented in the table.

An obvious generalization of velocity forcing involves systems of the form

$$\dot{x} = g(x, y) + A\sin\Omega t$$
$$\dot{y} = f(x, y) \tag{2.16}$$

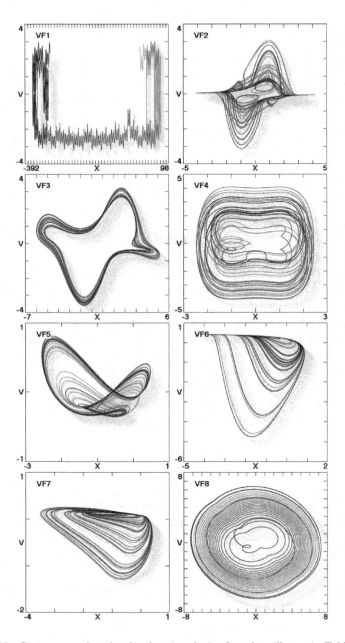

Fig. 2.12   State space plots for the chaotic velocity forced oscillators in Table 2.3.

Table 2.4   Chaotic nonlinear parametric oscillators.

Model	Equation	x_0, v_0, t_0	Lyapunov Exponents
PO_1	$\ddot{x} = \sin t \sin x$	$0, 1, 0$	$0.1717, 0, -0.1717$
PO_2	$\ddot{x} = (1 - x^2)\dot{x} - x \sin t$	$0.3, 0, 0$	$0.0620, 0, -4.4437$
PO_3	$\ddot{x} = (1 - \dot{x}^2)\dot{x} - 2 \sin t$	$1, 1, 0$	$0.0514, 0, -5.4477$
PO_4	$\ddot{x} = (\sin t - 1.2)x^3$	$5, 0, 0$	$0.0667, 0, -0.0667$
PO_5	$\ddot{x} = \dot{x} \sin t - x^3$	$1, 0, 0$	$0.0641, 0, -0.0641$
PO_6	$\ddot{x} = (x + \dot{x})x \sin t - 1.2x$	$-4, -5, 0$	$0.1308, 0, -2.3004$
PO_7	$\ddot{x} = (x - \dot{x}) \sin t - (x + 1)\dot{x}$	$1, 0.5, 0$	$0.0760, 0, -1.5178$
PO_8	$\ddot{x} = (1 - 3 \sin t)x^2 - (3\dot{x} + 1)x$	$0.3, 0, 0$	$0.1430, 0, -2.3343$

where f and g depend on both x and y. Such systems will not be explored here since they are less elegant than the more restricted forms already considered.

2.11   Parametric Oscillators

Although the usual case is for the forcing to act directly on either the acceleration or the velocity, it can also act on one or more parameters of the system. Such cases are called *parametric oscillators* and were first studied by Lord Rayleigh (Strutt, 1883, 1887). The usual example is a child on a swing in which energy is put into the system by modulating the natural frequency of the system by moving the center of mass up and down. Although a swing can certainly be pumped in this way, it is not a very good description of how most people actually swing (Case and Swanson, 1990; Case, 1996). The energy is transferred most efficiently if the forcing frequency Ω is twice the natural frequency of the system, and the equation describing such a system is

$$\ddot{x} + (1 + A \sin \Omega t)x = 0, \qquad (2.17)$$

which is a form of Mathieu's equation (Ruby, 1996). It can be shown that any such *linear* oscillator can only have solutions that are periodic or unbounded, and in particular, they cannot exhibit chaos.

However, *nonlinear* parametric oscillators can easily produce chaos, some elegant examples of which are given in Table 2.4 with their corresponding state space plots in Fig. 2.13.

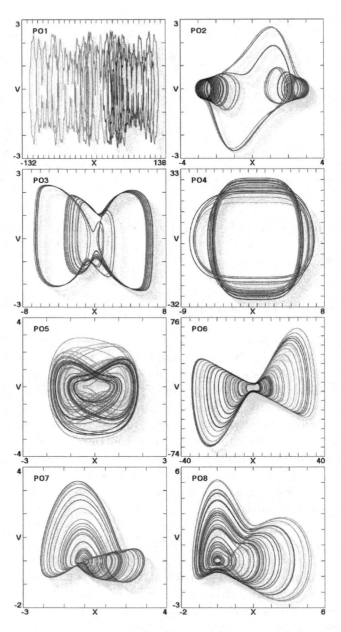

Fig. 2.13   State space plots for the chaotic nonlinear parametric oscillators in Table 2.4.

Table 2.5   Chaotic forced complex oscillators.

Model	Equation	z_0, t_0	Lyapunov Exponents
FZ_1	$\dot{z} + z^2 - \overline{z} + 1 = e^{it}$	$-0.8i, 0$	$0.0473, 0, -0.9869$
FZ_2	$\dot{z} + (z - \overline{z})z + 1 = e^{it}$	$0.7i, 0$	$0.0641, 0, -0.2031$
FZ_3	$\dot{z} + 2z^2 - \overline{z}^2 + 2 = e^{it}$	$0, 0$	$0.0803, 0, -0.0805$
FZ_4	$\dot{z} + 0.3z^3 + \overline{z} + 0.3 = e^{it}$	$-1 - 0.8i, 0$	$0.1231, 0, -0.6053$
FZ_5	$\dot{z} + (0.2z^2 + 1)z + \overline{z} = e^{it}$	$2 - 3i, 0$	$0.1540, 0, -1.3517$
FZ_6	$\dot{z} + (z^2 - \overline{z}^2)z + \overline{z} = e^{it}$	$0.6, 0$	$0.0892, 0, -0.1621$
FZ_7	$\dot{z} + (z^2 + z\overline{z} + \overline{z}^2 + 1)\overline{z} = e^{it}$	$0.5, 0$	$0.1157, 0, -6.1517$

2.12   Complex Oscillators

If we allow the variables to be complex numbers (with real and imaginary parts), that allows for the possibility of elegant chaotic systems of the form

$$\dot{z} + f(z, \overline{z}) = e^{i\Omega t} \tag{2.18}$$

where z is given in terms of the real variables x and y by $z = x + iy$ and its complex conjugate $\overline{z} = x - iy$ with $i^2 \equiv -1$. Note that z is a vector in the complex xy-plane rather than the phase of the forcing function as it was in the previous sections. If $f(z, \overline{z})$ depends only on z or only on $\overline{z}$, no chaotic solutions are possible (Marshall and Sprott, 2009). Equation (2.17) is just a shorthand way of writing

$$\begin{aligned} \dot{x} + \Re f(z, \overline{z}) &= \cos \Omega t \\ \dot{y} + \Im f(z, \overline{z}) &= \sin \Omega t \end{aligned} \tag{2.19}$$

where $\Re f$ and $\Im f$ are the real and imaginary parts of f, respectively.

With both z and $\overline{z}$ present, chaos occurs in three simple quadratic examples given by FZ_1 through FZ_3 in Table 2.5 and in four cubic examples given by FZ_4 through FZ_7 in the same table, with attractors as shown in Fig. 2.14. The model FZ_3 is only very weakly dissipative as evidenced by the Lyapunov exponents that almost sum to zero, but the others are strongly dissipative.

Of course complex oscillators can also be parametrically forced, one simple example of which is given by

$$\dot{z} = a(1 + e^{it}|z|^2)\overline{z}. \tag{2.20}$$

Srzednicki and Wójcik (1997) proved that this system is chaotic for $a > 288$. The calculation is numerically difficult for such a case because of the very different time scales that describe the fast (natural frequency) and

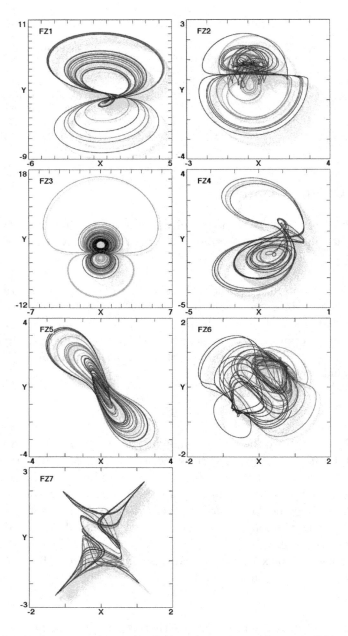

Fig. 2.14   Strange attractors for the forced complex oscillators in Table 2.5.

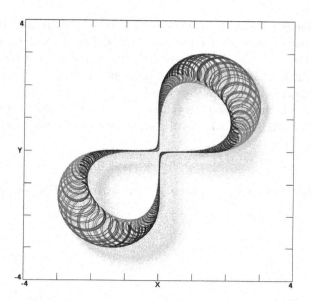

Fig. 2.15   State space plot for the parametrically forced complex oscillator in Eq. (2.20) with $a = 0.02$ and $(z_0, t_0) = (0.33 + 0.71i, 0)$, $\lambda = (0.0022, 0. - 0.0022)$.

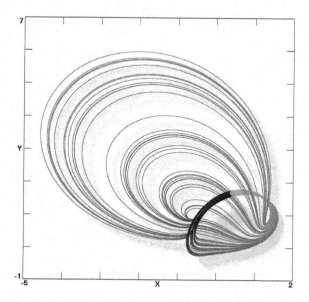

Fig. 2.16   Attractor for the parametrically forced complex oscillator in Eq. (2.21) with $(z_0, t_0) = (0.4 + 1.6i, 0)$, $\lambda = (0.0490, 0. - 0.6182)$.

the slow (forcing frequency) dynamics. However, there are also weakly chaotic solutions for small a, one example of which for $a = 0.02$ is shown in Fig. 2.15. This is a conservative system with two somewhat different time scales for which initial conditions much be chosen very carefully to achieve bounded chaotic solutions in the chaotic sea.

Other simple chaotic systems can be constructed by parametrically forcing the systems in Table 2.5, one of the simplest examples of which is given by

$$\dot{z} + z^2 e^{it} + \overline{z} + 1 = 0. \tag{2.21}$$

This is a well-behaved dissipative system with an attractor as shown in Fig. 2.16.

Chapter 3

Autonomous Dissipative Systems

The simplest and most common examples of chaos in continuous-time systems come from three first-order ordinary differential equations with no explicit time dependence. Such equations are said to be *autonomous*. These systems typically have dissipation and thus produce strange attractors in the three-dimensional state space of their variables.

3.1 Lorenz System

The story is now well known (Gleick, 1987; Lorenz, 1993) how meteorologist Edward Lorenz at the Massachusetts Institute of Technology in 1959, to whom this book is dedicated, accidentally discovered sensitive dependence on initial conditions while using a twelve-dimensional system of differential equations to model atmospheric convection on a primitive, 800-pound, desk-size digital computer (Royal McBee LGP-30) when he repeated his calculation with the six-digit initial conditions rounded to three digits, naively assuming that a difference of one part in a thousand would be inconsequential. He reported his results, along with their implications for weather prediction, at a meteorology conference in Tokyo in 1960, but meteorologists were slow to accept his conclusions because his model was deemed too simple. Undeterred, he subsequently simplified a seven-dimensional system with thirteen quadratic nonlinearities (Saltzman, 1962) by methods similar to those used throughout this book to what for its time was a very elegant three-dimensional autonomous dissipative system with two quadratic nonlinearities (Lorenz, 1963) given by

$$\dot{x} = \sigma(y - x)$$
$$\dot{y} = -xz + rx - y \qquad (3.1)$$
$$\dot{z} = xy - bz$$

61

with chaotic solutions for $\sigma = 10, r = 28$, and $b = 8/3$, and Lyapunov exponents $\lambda = (0.9056, 0, -14.5723)$.

This system has been widely studied, and there is a whole book devoted to it (Sparrow, 1982), but it took thirty-six years to prove that the system is chaotic (Tucker, 1999), dispelling a lingering fear that the chaos might be a numerical artifact (Stewart, 2000). The Lorenz equations well approximate several other chaotic systems, including a dynamo (Cook and Roberts, 1970), a laser (Weiss and Brock, 1986), and a waterwheel (Strogatz, 1994).

The fact that there are three parameters (σ, r, b) is no coincidence. One can linearly rescale the variables x, y, z, and t so that four of the seven terms on the right-hand side of Eq. (3.1) have coefficients of 1.0, leaving three terms with adjustable coefficients. The choice of where to put the remaining coefficients is arbitrary, although in this case, Lorenz chose them to represent physically relevant quantities. The parameters σ and r are proportional to the Prandtl number and the Rayleigh number, respectively, and the parameter b is the aspect ratio of the convection cylinders.

The Lorenz system can be made marginally more elegant by using the parameters $\sigma = 4, r = 16$, and $b = 1$, for which the attractor shown in Fig. 3.1 closely resembles the familiar one. This and other similar figures throughout the book show only a portion of a typical three-dimensional trajectory on the attractor projected onto a plane of two of the variables (x and y in this case) with the third variable (z in this case) displayed in shades of gray to indicate its height above the plane with a subtle shadow below and to the right to give an illusion of depth. Initial conditions (x_0, y_0, z_0) are given that are close to the attractor and certainly within its basin of attraction, which for the Lorenz system is the entire state space except for the equilibrium points at the origin and at the centers of the two lobes.

Figure 3.2 shows in the space of the remaining two parameters (σ and r), the regions of different dynamical behaviors, with black indicating chaos (C) and dark gray indicating a stable equilibrium (S) as determined by the sign of the largest Lyapunov exponent. The horizontal light-gray line at $r = 1$ represents a region where $|\lambda| < 0.001$, and it is a bifurcation boundary where the origin $(0, 0, 0)$ becomes unstable and two stable equilibria are born at $x = y = \pm\sqrt{b(r - 1)}$ and $z = r - 1$, which become the centers of the two lobes of the strange attractor.

Chaotic solutions are also obtained with the parameter σ set to 1.0, for example at $(\sigma, r, b) = (1, 16, 0.03)$ with initial conditions $(x_0, y_0, z_0) = (0.3, 0.5, 13)$ and $\lambda = (0.0531, 0, -2.0831)$. If we allow negative parameters, it is possible to set both σ and b to unity, for example at

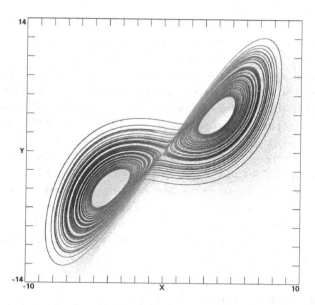

Fig. 3.1 Lorenz attractor from Eq. (3.1) for $(\sigma, r, b) = (4, 16, 1)$ with initial conditions $(x_0, y_0, z_0) = (1, 2, 6)$, $\lambda = (0.3359, 0, -6.3359)$.

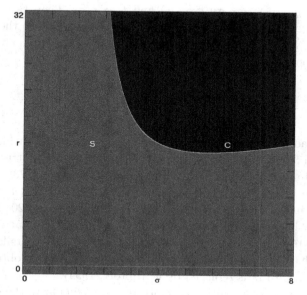

Fig. 3.2 Regions of different dynamics for the Lorenz attractor from Eq. (3.1) with $b = 1$.

$(\sigma, r, b) = (1, -17, -1)$ with initial conditions $(x_0, y_0, z_0) = (8.5, 13, 0.4)$ and $\lambda = (0.0629, 0, -1.0629)$. The resulting system then has a single parameter (r) for which chaos occurs over the range $-17.025 < r < -16.220$.

The Lorenz system can be made even more elegant by setting the $-y$ term in the $\dot{y}$ equation to zero (Zhou *et al.*, 2008), which gives chaos for $(\sigma, r, b) = (1, 2, 0.1)$ with initial conditions $(x_0, y_0, z_0) = (1, 0, 3)$ and $\lambda = (0.0413, 0, -1.1413)$ or by reversing its sign (Lü *et al.*, 2002), which gives chaos for $(\sigma, r, b) = (2, 0, 0.2)$ with initial conditions $(x_0, y_0, z_0) = (1, 0, 2)$ and $\lambda = (0.0599, 0, -1.2599)$. These appear to be the only terms that can be removed from the Lorenz system without destroying the chaos. Thus only six of the seven terms and two of the three parameters in the usual form of the Lorenz system are required to get a strange attractor resembling the one found by Lorenz. The transition between the standard Lorenz attractor and the ones with the x and y terms removed from the $\dot{y}$ equation has been studied by Sun and Sprott (2009).

3.2   Diffusionless Lorenz System

Although the Lorenz system cannot be further simplified by taking any of the other terms to zero without losing the chaos, it can be rescaled as follows: $(x, y, z) \to (\sigma x, \sigma y, \sigma z + r)$ and $t \to t/\sigma$. Then take $r, \sigma \to \infty$ but in such a way that $R = br/\sigma^2$ remains finite, leading to the *diffusionless Lorenz system* (van der Schrier and Maas, 2000; Munmuangsaen and Srisuchinwong, 2009)

$$
\begin{aligned}
\dot{x} &= y - x \\
\dot{y} &= -xz \\
\dot{z} &= xy - R,
\end{aligned}
\tag{3.2}
$$

which is chaotic for a wide range of the single parameter R including $R = 1$ as shown in Fig. 3.3. Its attractor has the familiar two-lobe structure of the Lorenz system, but with a higher dimension of $D_{KY} = 2.1736$ in contrast to the case in Fig. 3.1 whose dimension is $D_{KY} = 2.053$ and the case Lorenz studied whose dimension is $D_{KY} = 2.062$ (Sprott, 2003). In fact, this system has its maximum dimension of $D_{KY} = 2.2354$ at $R = 3.4693$ (Sprott, 2007a). Its route to chaos is shown in Fig. 3.4. A similar two-lobe attractor that also resembles the familiar Lorenz attractor is obtained if the xy term in Eq. (3.2) is replaced by y^2 (Sprott, 1994).

Many other systems with a two-lobe topology similar to the Lorenz system have been proposed and studied including ones with a pair of quadratic

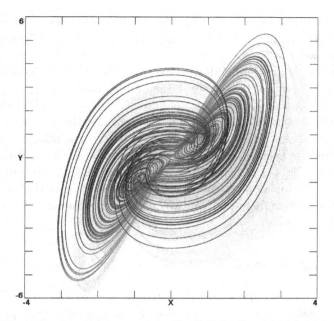

Fig. 3.3 Diffusionless Lorenz attractor from Eq. (3.2) for $R = 1$ with initial conditions $(x_0, y_0, z_0) = (1, 0, 1)$, $\lambda = (0.2101, 0, -1.2101)$.

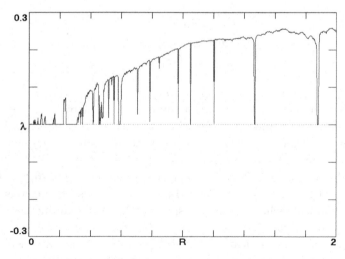

Fig. 3.4 Regions of different dynamics for the diffusionless Lorenz attractor from Eq. (3.2) with initial conditions $(x_0, y_0, z_0) = (1, 0, 1)$.

nonlinearities as suggested by Shimizu and Moroika (1980), Shaw (1981), Vallis (1988), Rucklidge (1992), Wang *et al.* (1992), and Chen and Ueta (1999), as well as ones with more than two quadratic nonlinearities as suggested by Rikitake (1958) (see later in this chapter), Moffat (1979), Leipnik and Newton (1981), Zeghlache and Mandel (1985), and Liu and Chen (2003), and ones with cubic nonlinearities as suggested by Moore and Spiegel (1966), Arneodo *et al.* (1981), Auvergne and Baglin (1985), Chua *et al.* (1986), Thomas (1999), and Qi *et al.* (2005), exponential nonlinearities as suggested by Duan *et al.* (2005), and piecewise-linear nonlinearities as suggested by Chua *et al.* (1986) and Elwakil *et al* (2002). The latter case is especially nice, but none of these systems is as elegant as the diffusionless Lorenz system with its five terms, two quadratic nonlinearities, and a single adjustable parameter.

3.3   Rössler System

Although the diffusionless Lorenz system is a considerable simplification of the original Lorenz system, its elegance is marred by the presence of two nonlinearities, which is one more than needed for chaos. To overcome that problem, Otto Rössler (1976), a nonpracticing medical doctor, cleverly concocted a simpler system with a single quadratic nonlinearity given by

$$\dot{x} = -y - z$$
$$\dot{y} = x + ay \qquad\qquad (3.3)$$
$$\dot{z} = b + z(x - c),$$

which is chaotic for $a = b = 0.2, c = 5.7$.

This system can be made marginally more elegant by using the parameters $a = 0.5, b = 1$, and $c = 3$, for which the attractor is shown in Fig. 3.5. Its dimension $D_{KY} = 2.0605$ is greater than the value of $D_{KY} = 2.0132$ for the parameters used by Rössler (Sprott, 2003), but less than the maximum value of $D_{KY} = 2.1587$, which occurs for $a = 0.6276, b = 0.7980$, and $c = 2.0104$ (Sprott, 2007a). Figure 3.6 shows the regions of different dynamical behaviors in the space of the two parameters a and c for $b = 1$, with black indicating chaos (C), light gray indicating periodicity (P), dark gray indicating a stable equilibrium (S), and white indicating unbounded solutions (U) as determined by the sign of the largest Lyapunov exponent.

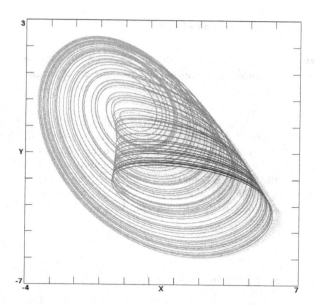

Fig. 3.5 Rössler attractor from Eq. (3.3) for $(a, b, c) = (0.5, 1, 3)$ with initial conditions $(x_0, y_0, z_0) = (2, 0, 1)$, $\lambda = (0.1172, 0, -1.9383)$.

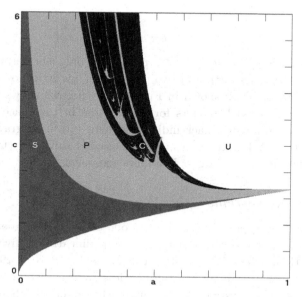

Fig. 3.6 Regions of different dynamics for the Rössler attractor from Eq. (3.3) with $b = 1$.

3.4   Other Quadratic Systems

For many years the Lorenz and Rössler systems were regarded as the simplest examples of chaos in autonomous dissipative systems of ODEs. In fact, Lorenz (1993) wrote:

> One other study left me with mixed feelings. Otto Rössler of the University of Tübingen had formulated a system of three differential equations as a model of a chemical reaction. By this time, a number of systems of differential equations with chaotic solutions had been discovered, but I felt I still had the distinction of having found the simplest. Rössler changed things by coming along with an even simpler one. His record still stands.

3.4.1   *Rössler prototype-4 system*

What Lorenz apparently did not realize was that Rössler himself had much earlier found an even simpler system (Rössler, 1979a) given by

$$\dot{x} = -y - z$$
$$\dot{y} = x \tag{3.4}$$
$$\dot{z} = a(y - y^2) - bz.$$

This system (called *Rössler prototype-4*) has only six terms, a single quadratic nonlinearity (y^2), and two parameters, giving chaos for $a = b = 0.5$ with an attractor as shown in Fig. 3.7. Figure 3.8 shows the regions of different dynamical behaviors for this system in the space of the two parameters a and b, with black indicating chaos (C), light gray indicating periodicity (P), and white indicating unbounded solutions (U) as determined by the sign of the largest Lyapunov exponent.

3.4.2   *Sprott systems*

The Rössler prototype-4 system is only one of a large number of chaotic systems with six terms and one quadratic nonlinearity or five terms and two quadratic nonlinearities as shown in Table 3.1 (Sprott, 1994). Model SQ_B is in fact a variant of the diffusionless Lorenz system in Eq. (3.2), but all 18 of these systems were apparently unknown when they were first discovered through an extensive numerical search. The attractors

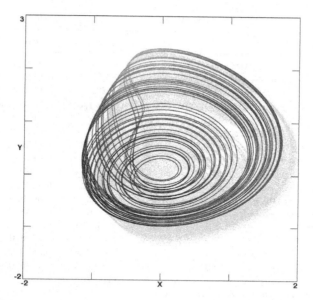

Fig. 3.7 Rössler prototype-4 attractor from Eq. (3.4) for $a = b = 0.5$ with initial conditions $(x_0, y_0, z_0) = (0.1, 0.3, 0)$, $\lambda = (0.0938, 0, -0.5938)$.

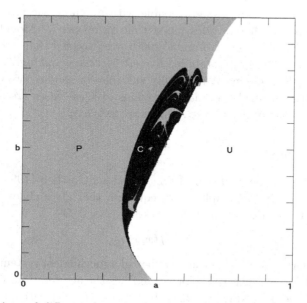

Fig. 3.8 Regions of different dynamics for the Rössler prototype-4 attractor from Eq. (3.4).

Table 3.1 Simple three-dimensional chaotic flows with quadratic nonlinearities.

Model	Equations	x_0, y_0, z_0	Lyapunov Exponents
SQ_B	$\dot{x} = yz, \dot{y} = x - y, \dot{z} = 1 - xy$	$0.6, 0, 0$	$0.2101, 0, -1.2101$
SQ_C	$\dot{x} = yz, \dot{y} = x - y, \dot{z} = 1 - x^2$	$0, 0.5, 0$	$0.1633, 0, -1.1633$
SQ_D	$\dot{x} = -y, \dot{y} = x + z, \dot{z} = xz + 3y^2$	$-0.2, 0, 1$	$0.1027, 0, -1.3198$
SQ_E	$\dot{x} = yz, \dot{y} = x^2 - y, \dot{z} = 1 - 4x$	$-1, 1, 0$	$0.0774, 0, -1.0774$
SQ_F	$\dot{x} = y + z, \dot{y} = -x + 0.5y, \dot{z} = x^2 - z$	$0, 0.2, 0$	$0.1171, 0, -0.6171$
SQ_G	$\dot{x} = 0.4x + z, \dot{y} = xz - y, \dot{z} = -x + y$	$1, 0, 0$	$0.0340, 0, -0.6340$
SQ_H	$\dot{x} = -y + z^2, \dot{y} = x + 0.5y, \dot{z} = x - z$	$0.1, 0, 0$	$0.1171, 0, -0.6171$
SQ_I	$\dot{x} = -0.2y, \dot{y} = x + z, \dot{z} = x + y^2 - z$	$0, 0.3, 0$	$0.0125, 0, -1.0125$
SQ_J	$\dot{x} = 2z, \dot{y} = -2y + z, \dot{z} = -x + y + y^2$	$0, 1, 4$	$0.0757, 0, -2.0757$
SQ_K	$\dot{x} = xy - z, \dot{y} = x - y, \dot{z} = x + 0.3z$	$1, 1, 2$	$0.0375, 0, -0.8900$
SQ_L	$\dot{x} = y + 3.9z, \dot{y} = 0.9x^2 - y, \dot{z} = 1 - x$	$0, 12, -6$	$0.0609, 0, -1.0609$
SQ_M	$\dot{x} = -z, \dot{y} = -x^2 - y, \dot{z} = 1.7 + 1.7x + y$	$1, -0.8, 0$	$0.0435, 0, -1.0435$
SQ_N	$\dot{x} = -2y, \dot{y} = x + z^2, \dot{z} = 1 + y - 2z$	$4.5, 1, 0$	$0.0758, 0, -2.0758$
SQ_O	$\dot{x} = y, \dot{y} = x - z, \dot{z} = x + xz + 2.7y$	$0, 0, 0.5$	$0.0488, 0, -0.3187$
SQ_P	$\dot{x} = 2.7y + z, \dot{y} = -x + y^2, \dot{z} = x + y$	$0, 0.3, 0$	$0.0873, 0, -0.4814$
SQ_Q	$\dot{x} = -z, \dot{y} = x - y, \dot{z} = 3.1x + y^2 + 0.5z$	$1, 0, 0$	$0.1085, 0, -0.6085$
SQ_R	$\dot{x} = 0.9 - y, \dot{y} = 0.4 + z, \dot{z} = xy - z$	$2, 0, 0$	$0.0622, 0, -1.0622$
SQ_S	$\dot{x} = -x - 4y, \dot{y} = x + z^2, \dot{z} = 1 + x$	$0, 0, 1$	$0.1876, 0, -1.1876$

various cases are shown in Fig. 3.9. Case SQ_D has the unusual property of being dissipative but time-reversible with a symmetric pair of strange attractors that exchange roles when time is reversed. Model SQ_E has been studied by González-Miranda (2006), who showed that it exhibits unusual behavior including bistability, hysteresis, crises, and highly incoherent phase dynamics, making it especially suitable for studies of synchronization and chaos control. Animated versions of these cases are available at http://sprott.physics.wisc.edu/simplest.htm.

3.5 Jerk Systems

Confronted with the results in Table 3.1, Hans Gottlieb (1996) posed the question 'What is the simplest jerk equation that gives chaos?' by which he meant an equation of the form

$$\dddot{x} = f(\ddot{x}, \dot{x}, x). \tag{3.5}$$

The term 'jerk' comes from the fact that in a mechanical system in which x is the displacement, $\dot{x}$ is the velocity, and $\ddot{x}$ is the acceleration, the quantity $\dddot{x}$ is called the 'jerk' (Schot, 1978). It is the lowest derivative for which an ODE with smooth continuous functions can give chaos.

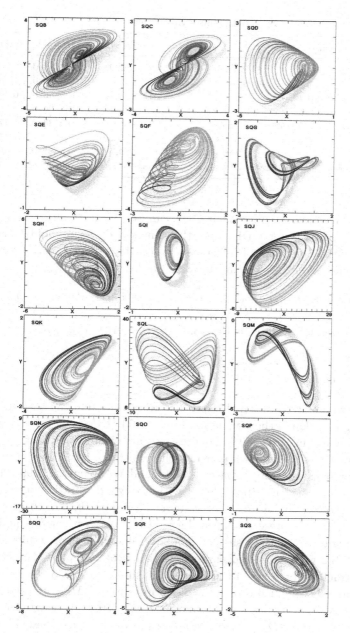

Fig. 3.9   Attractors from the simple chaotic quadratic systems in Table 3.1.

Table 3.2   Simple chaotic jerk systems.

Model	Equation	$x_0, \dot{x}_0, \ddot{x}_0$	Lyapunov Exponents
JD_0	$\dddot{x} = -2.02\ddot{x} + \dot{x}^2 - x$	$5, 2, 0$	$0.0486, 0, -2.0686$
JD_1	$\dddot{x} = -1.8\ddot{x} - 2x + x\dot{x} - 1$	$-5, 5, 0.9$	$0.0647, 0, -1.8647$
JD_2	$\dddot{x} = -0.4\ddot{x} - 2.1\dot{x} + x^2 - 1$	$0, 2, -1$	$0.0856, 0, -0.4856$
JD_3	$\dddot{x} = -0.6\ddot{x} - \dot{x} + 0.9x^2 + x\dot{x} - 1$	$-2, 0, 3$	$0.0809, 0, -0.6809$
JD_4	$\dddot{x} = -\ddot{x} - 0.7\dot{x} + x^2 + x\ddot{x} - 1$	$-2, 0, 1$	$0.0734, 0, -1.7769$
JD_5	$\dddot{x} = -\dot{x} + 3\dot{x}^2 - x^2 - x\ddot{x}$	$1, 0.2, 0$	$0.1028, 0, -1.3198$
JD_6	$\dddot{x} = -\ddot{x} - \dot{x} + 2x^2 + 2\dot{x}^2 + x\ddot{x} - 1$	$0, 0, -0.7$	$0.0655, 0, -1.4662$
JD_7	$\dddot{x} = -\ddot{x} + \dot{x} + 2x^2 - 3\dot{x}^2 + x\dot{x} + x\ddot{x} - 1$	$-0.7, 0, 0$	$0.0771, 0, -3.1105$

It is known that any explicit ODE can be cast in the form of a system of coupled first-order ODEs, but the converse does not hold in general. However, the Lorenz and Rössler systems can be written in jerk form (Linz, 1997). The Lorenz system in Eq. (3.1) can be written as

$$\dddot{x} + (1 + \sigma + b - \dot{x}/x)\ddot{x} + [b(1 + \sigma + x^2) - (1 + \sigma)\dot{x}/x]\dot{x} - b\sigma(r - 1 - x^2)x = 0, \tag{3.6}$$

and the Rössler system in Eq. (3.3) can be written as (Innocenti *et al.*, 2008)

$$\dddot{y} + (c - a)\ddot{y} + (1 - ac)\dot{y} + cy - b - (\dot{y} - ay)(\ddot{y} - a\dot{y} + y) = 0, \tag{3.7}$$

both of which are rather inelegant. The Rössler system can also be written as a jerk function of the x variable (Linz, 1997) or the z variable (Lainscsek *et al.*, 2003), but the resulting expressions are even less elegant.

However, in a landmark paper, Eichhorn *et al.* (1998) used the method of comprehensive Gröbner bases (Becker and Weispfenning, 1993) to show that all 14 of the cases in Table 3.1 with a single quadratic nonlinearity as well as the Rössler prototype-4 system and model SQ_D with two nonlinearities can be represented in an hierarchy of quadratic jerk equations of increasing complexity. A slightly modified version of their results is shown in Table 3.2 with simplified parameters that produce chaos (Sprott, 2003). The attractors for some of these cases are shown in Fig. 3.10. Model JD_0 and model JD_2, which is of the same form as model MO_3 in Table 3.3 will be discussed later. See also Eichhorn *et al.* (2002) for a detailed discussion of JD_1 and JD_2. All of these cases have the same 'folded-band' (or 'one-scroll') structure as the Rössler system and are thus the topologically simplest examples of chaos.

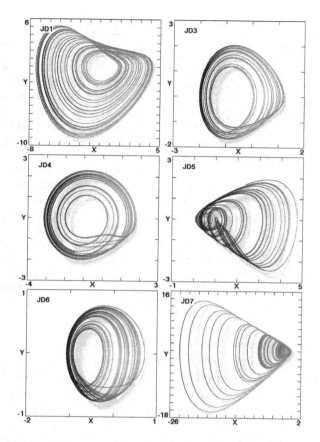

Fig. 3.10 Attractors from some of the simple chaotic jerk systems in Table 3.2.

3.5.1 *Simplest quadratic case*

The simplest case (JD_0) in the hierarchy of quadratic jerk equations is (Sprott, 1997a)

$$\dddot{x} = -a\ddot{x} + \dot{x}^2 - x, \tag{3.8}$$

which is chaotic for $a = 2.02$. Its attractor is shown in Fig. 3.11, and the scaling of its largest Lyapunov exponent with the single parameter a is shown in Fig. 3.12. An animated version of this case along with its basin of attraction is available at `http://sprott.physics.wisc.edu/simpjerk.htm`. If one accepts that the quadratic is the simplest nonlinearity, and the

Table 3.3 Chaotic memory oscillators.

Model	Equation	$x_0, \dot{x}_0, \ddot{x}_0$	Lyapunov Exponents
MO_0	$\dddot{x} + 0.6\ddot{x} + \dot{x} = \lvert x \rvert - 1$	$0, -0.7, 0$	$0.0363, 0, -0.6363$
MO_1	$\dddot{x} + 0.6\ddot{x} + \dot{x} = 1 - 6\max(x, 0)$	$0, 0, 0.5$	$0.0926, 0, -0.6926$
MO_2	$\dddot{x} + 0.6\ddot{x} + \dot{x} = \operatorname{sgn} x - x$	$0, 0, 1$	$0.1416, 0, -0.7416$
MO_3	$\dddot{x} + \ddot{x} + \dot{x} = 1.1(x^2 - 1)$	$1, 0, -1$	$0.0574, 0, -1.0574$
MO_4	$\dddot{x} + 0.5\ddot{x} + \dot{x} = x(x - 1)$	$0, 0.1, 0$	$0.0938, 0, -0.5938$
MO_5	$\dddot{x} + 0.7\ddot{x} + \dot{x} = x(1 - x^2)$	$0, 0, 0.1$	$0.1380, 0, -0.8380$
MO_6	$\dddot{x} + 0.4\ddot{x} + \dot{x} = x^2(1 - x)$	$0, 0, 0.5$	$0.0245, 0, -0.4245$
MO_7	$\dddot{x} + 0.6\ddot{x} + \dot{x} = x^2(1 - x^2)$	$0, 0, 0.4$	$0.0567, 0, -0.6567$
MO_8	$\dddot{x} + 0.5\ddot{x} + \dot{x} = x(x^4 - 1)$	$0, 0.9, 0$	$0.0786, 0, -0.5786$
MO_9	$\dddot{x} + 0.4\ddot{x} + \dot{x} = x^3(1 - x)$	$1, 0.2, 0$	$0.0537, 0, -0.4537$
MO_{10}	$\dddot{x} + 0.6\ddot{x} + \dot{x} = x^2(1 - x^3)$	$0, 0, 0.4$	$0.0315, 0, -0.6315$
MO_{11}	$\dddot{x} + \ddot{x} + \dot{x} = 5 - e^x$	$0, 4.3, 0$	$0.0352, 0, -1.0352$
MO_{12}	$\dddot{x} + 0.5\ddot{x} + \dot{x} = 7 - 8\tanh x$	$0, 1, 6$	$0.0254, 0, -0.5254$
MO_{13}	$\dddot{x} + \ddot{x} + \dot{x} = 6\tanh x - 3x$	$0, 0, 1$	$0.0970, 0, -1.0970$
MO_{14}	$\dddot{x} + 0.6\ddot{x} + \dot{x} = 6\arctan x - x$	$0, 1, 6$	$0.0812, 0, -0.6812$
MO_{15}	$\dddot{x} + 0.6\ddot{x} + \dot{x} = x - 0.5\sinh x$	$0, 0, 1$	$0.0638, 0, -0.6638$

third is the lowest derivative for which chaos occurs, then Eq. (3.8) must be the algebraically simplest continuous-time chaotic system. Any polynomial system with fewer terms would have no adjustable parameters, severely limiting the range of its possible dynamics. In fact, it has been rigorously proved that there can be no simpler system that is chaotic (Zhang and Heidel, 1997, 1999).

Equation (3.8) can be written in equivalent forms by replacing x with $-x$, which changes the sign of the quadratic term and yields an attractor that is the mirror image of the one in Fig. 3.11, or by differentiating each term with respect to time and defining $y \equiv 2\dot{x}$ to get

$$\dddot{y} = -a\ddot{y} + y\dot{y} - y. \tag{3.9}$$

Equation (3.9) is a special case of model JD_1 in Table 3.2 with the constant term (-1) equal to zero, and cases SQ_I, SQ_J, SQ_L, and SQ_R in Table 3.1 are also of that form (Eichhorn *et al.*, 1998). Letellier and Valée (2003) studied Eq. (3.8) and (3.9) in detail, and Malasoma (2002) showed that these jerk forms can arise from six different three-dimensional systems with only five terms and a single quadratic nonlinearity, which he called 'class P' systems.

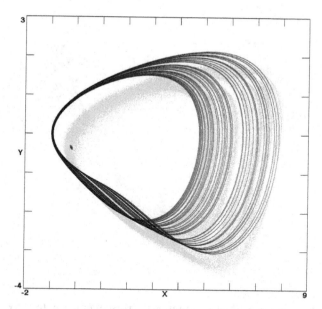

Fig. 3.11 Attractor for the simplest chaotic jerk system from Eq. (3.8) for $a = 2.02$ with initial conditions $(x_0, y_0, z_0) = (4, 2, 0)$, $\lambda = (0.0486, 0, -2.0686)$.

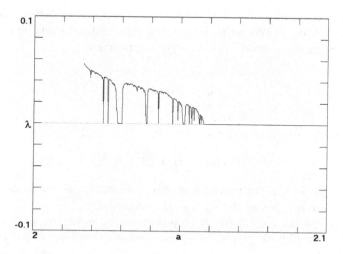

Fig. 3.12 Regions of different dynamics for the simplest chaotic jerk system from Eq. (3.8) with initial conditions $(x_0, y_0, z_0) = (4, 2, 0)$.

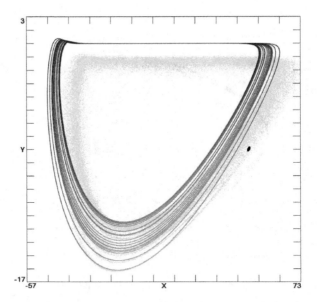

Fig. 3.13   Attractor for the Malasoma rational jerk system in Eq. (3.10) with $\alpha = 10.3$ and initial conditions $(x_0, y_0, z_0) = (46, 1, 10)$, $\lambda = (0.0350, 0, -10.3350)$.

3.5.2   *Rational jerks*

Malasoma (2002) also described three chaotic systems with five terms and a single quadratic nonlinearity whose jerk representation involves rational functions with four terms. One such system is given by

$$\dot{x} = z$$
$$\dot{y} = -\alpha y + z \qquad\qquad (3.10)$$
$$\dot{z} = -x + xy,$$

which can be written in jerk form as

$$\dddot{x} = -\alpha x - \alpha \ddot{x} + x\dot{x} + \dot{x}\ddot{x}/x. \qquad (3.11)$$

This example illustrates that the jerk representation is not necessarily a simplification, at least in the form of the nonlinearity.

The system in Eq. (3.10) is chaotic over most of the narrow range $10.2849 < \alpha < 10.3716$ with an attractor as shown in Fig. 3.13 for $\alpha = 10.3$. This system is very strongly damped, with an attractor that is extremely thin in the vicinity of $y = 1$ and $z = \alpha$ and almost planar, with a Kaplan–Yorke dimension of $D_{KY} = 2.0034$.

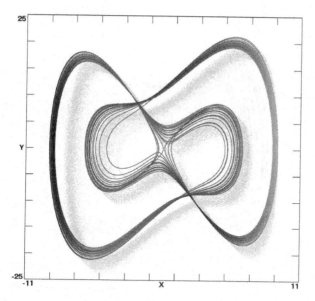

Fig. 3.14   Attractor for the Moore–Spiegel system from Eq. (3.12) with initial conditions $(x_0, y_0, z_0) = (4, 7, 2)$, $\lambda = (0.0652, 0, -1.0652)$.

3.5.3   *Cubic cases*

If one allows cubic nonlinearities, several simple cases are known (Sprott, 1997b) including a very old one found by Moore and Spiegel (1966) as a model for the irregular variability in the luminosity of stars and given in simplified form by

$$\dddot{x} = -\ddot{x} + 9\dot{x} - x^2\dot{x} - 5x \tag{3.12}$$

with an attractor as shown in Fig. 3.14. Moore and Spiegel noted the aperiodic behavior of their solutions and the sensitive dependence on initial conditions but were apparently unaware of Lorenz's similar work. Had they published a few years earlier and emphasized the implications of their results to predictability, they might have become as famous as Lorenz.

The simplest cubic case proposed by Malasoma (2000) is

$$\dddot{x} = -a\ddot{x} + x\dot{x}^2 - x, \tag{3.13}$$

which is chaotic for $a = 2.03$ with an attractor as shown in Fig. 3.15 and a route to chaos as shown in Fig. 3.16. Malasoma conjectures that this is the simplest system that is invariant under the parity transformation $x \to -x$. This case was further studied by Letellier and Valée (2003).

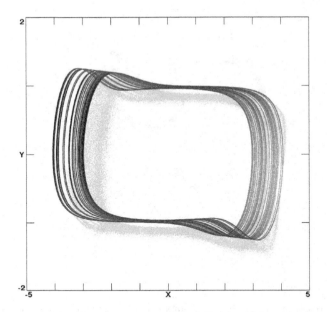

Fig. 3.15   Attractor for the cubic jerk system from Eq. (3.13) for $a = 2.03$ with initial
conditions $(x_0, y_0, z_0) = (0, 0.96, 0)$, $\lambda = (0.0769, 0, -2.1069)$.

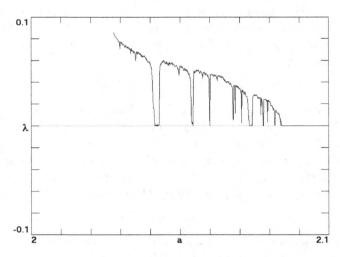

Fig. 3.16   Regions of different dynamics for the cubic jerk system from Eq. (3.13) with
initial conditions $(x_0, y_0, z_0) = (0, 0.96, 0)$.

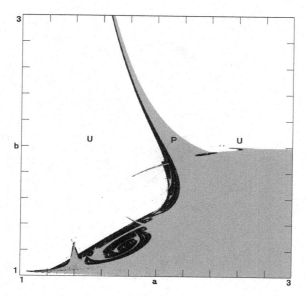

Fig. 3.17 Regions of different dynamics for the generalized jerk system from Eq. (3.15).

3.5.4 *Cases with arbitrary power*

The similarity of Eq. (3.9) and (3.13) suggests that there is a family of chaotic systems of the form

$$\dddot{x} = -a\ddot{x} + x|\dot{x}|^b - x, \tag{3.14}$$

and such appears to be the case, at least for $b > 0$.

This observation raises the question of whether there is an even simpler family of systems of the form

$$\dddot{x} = -a\ddot{x} + |\dot{x}|^b - x, \tag{3.15}$$

and in particular whether there is chaos for $b = 1$. The absolute value is needed to preserve the symmetry about $x = 0$ and to avoid imaginary numbers when b is not an integer. The chaos in this system occurs for all values of $b > 1$ as shown in Fig. 3.17. The spiral structure of the chaotic region and other features of the plot deserve further study. Note the counterintuitive result that increasing the nonlinearity (larger b) does not in general increase the region of parameter space over which chaos occurs.

For $b \to 1$, the attractor increases in size according to $2(\pi a^{b/2}/2)^{\frac{1}{b-1}}$, and this growth can cause numerical difficulties. However, the problem

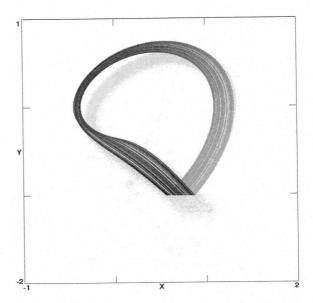

Fig. 3.18   Attractor for the square well system in Eq. (3.15) for $a = 0.816$ and $b = 1000$ with initial conditions $(x_0, y_0, z_0) = (0.5, 0.7, 0)$, $\lambda = (0.0748, 0, -0.8908)$.

is easily surmounted by including a coefficient $\pi a^{b/2}/2$ in the $|\dot{x}|^b$ term in Eq. (3.15), which has no effect other than to keep the size of the attractor of order unity as b approaches 1.0, and it allows one to determine that chaos does not exist for $b = 1$. For $b < 1$, the solution is unbounded

There is a narrow region of chaos that extends to arbitrarily large values of b. In the limit of $b \to \infty$, the nonlinear term resembles a square well with sides at $\dot{x} = \pm 1$, and the chaos arises from collisions with the side wall of the well at $\dot{x} = -1$, while the trajectory within the well is described by a linear system of equations. Figure 3.18 shows the strange attractor for the case of $a = 0.816$ and $b = 1000$. For this value of b, the nonlinear term $|\dot{x}|^b$ rises abruptly from 4.3×10^{-5} at $\dot{x} = -0.99$ to 2.1×10^4 at $\dot{x} = -1.01$, and the chaos is in a narrow band in the vicinity of $a = 0.816$.

3.5.5   *Piecewise-linear case*

The simplest chaotic system of finite size with an absolute value nonlinearity appears to be (Linz and Sprott, 1999)

$$\dddot{x} = -a\ddot{x} - \dot{x} + |x| - 1, \qquad (3.16)$$

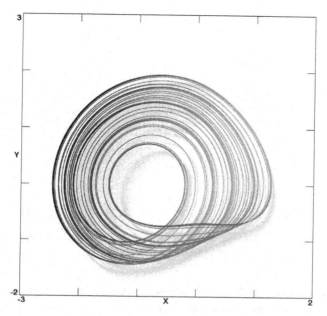

Fig. 3.19   Attractor for the piecewise-linear jerk system from Eq. (3.16) for $a = 0.6$ and $b = 1$ with initial conditions $(x_0, y_0, z_0) = (0, 1, -1)$, $\lambda = (0.0364, 0, -0.6364)$.

which is chaotic for $a = 0.6$ with an attractor as shown in Fig. 3.19. The scaling of its largest Lyapunov exponent with the parameter a is shown in Fig. 3.20. This system is a piecewise-linear variant of the quadratic system JD_2 in Table 3.2 and is arguably the simplest piecewise-linear chaotic flow. The abrupt change in direction of the flow at $x = 0$ is not evident in the Fig. 3.19 because the discontinuity occurs only in the fourth time derivative $\dddot{x}$ of x. The constant 1 in Eq. (3.16) only affects the size of the attractor. Chaos exists for arbitrarily small values of this constant, but the attractor and its basin of attraction shrink proportionally.

Linz (2000) has proved that chaos cannot exist in Eq. (3.16) if any of the terms are set to zero. He also notes that chaos is possible if the $|x|$ term is replaced with $|x^n|$, $|x|^n$, or x^{2n}, with n a positive integer, or more generally with any inversion-symmetric function $f(x) = f(-x)$ that is either positive or negative for all x. Numerical experiments indicate that chaotic solutions with $f(x) = |x|^n$ exist for all nonzero n, including noninteger and negative values. This system is one of a small number ODEs that can be rigorously shown to exhibit chaos (Zhezherun, 2005).

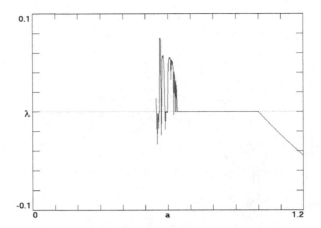

Fig. 3.20 Regions of different dynamics for the piecewise-linear jerk system from Eq. (3.16) with initial conditions $(x_0, y_0, z_0) = (0, 1, -1)$.

This particular nonlinearity is especially amenable to implementation with electronic circuits since the absolute value corresponds to a full-wave rectifier that can be achieved with diodes (Horowitz and Hill, 1989). A circuit that solves this equation will be described in Chapter 10.

3.5.6 *Memory oscillators*

Model JD_2 in Table 3.2 and Eq. (3.16) suggest that there might be a family of simple chaotic systems with the general form

$$\dddot{x} + a\ddot{x} + \dot{x} = g(x), \tag{3.17}$$

and Table 3.3 (Sprott, 2003) indicates that such is the case. Cases SQ_M, SQ_Q, and SQ_S in Table 3.1 as well as the Rössler prototype-4 system in Eq. (3.4) can all be reduced to that form (Eichhorn *et al.*, 1998). Model MO_0 is Eq. (3.16) with an attractor as shown in Fig. 3.19. Model MO_2 is the 'double scroll' system studied by Elwakil and Kennedy (2001) and implemented electronically. Model MO_3 is of the same form as model JD_2 in Table 3.2. Model MO_4 is an old case first noted by Coullet *et al.* (1979). Model MO_{13} turned up in a study of the interaction of the solar wind with the Earth's ionized magnetosphere (Horton *et al.*, 2001). Patidar and Sud (2005) have also investigated Eq. (3.17) with a variety of forms for $g(x)$ and concluded that while some nonlinearity is essential for chaos, more is not necessarily better.

Integrating each term in Eq. (3.17)

$$\ddot{x} + a\dot{x} + x = \int_{-\infty}^{t} g(x)dt \tag{3.18}$$

shows that the resulting system is a damped harmonic oscillator forced by a nonlinear memory term that involves a time integral over all past values of $x(t)$. Equation (3.17) may be a useful general model for fitting chaotic experimental data since the procedure reduces to finding a value of a and a function $g(x)$, or perhaps a more general function $g(\dot{x}, x)$, that gives the best fit to the data. It can be viewed as an extension of finding a potential $V(x)$ for a damped nonlinear oscillator governed by Newton's second law

$$\ddot{x} + a\dot{x} = -\frac{dV}{dx} \tag{3.19}$$

except that it permits chaotic solutions, which Eq. (3.19) does not. Constraining a to be positive ensures that the system is dissipative with an attractor for trajectories that lie within the basin of attraction.

Different forms of $g(x)$ produce a wide variety of attractors, some examples of which from Table 3.3 are shown in Fig. 3.21. Of course the methods are not mutually exclusive and can be combined in a system such as

$$\ddot{x} + a\dot{x} = -\frac{dV}{dx} + \eta$$
$$\dot{\eta} = g(\dot{x}, x), \tag{3.20}$$

but then the system begins to lose some of its elegance. Systems of this form with g dependent only on x have been studied by Arneodo *et al.* (1982).

3.6 Circulant Systems

A particularly elegant system is one in which the variables are cyclically symmetric according to

$$\dot{x} = f(x, y, z)$$
$$\dot{y} = f(y, z, x) \tag{3.21}$$
$$\dot{z} = f(z, x, y)$$

where all the functions are the same except the variables are rotated. A number of such circulant systems are known to produce chaos, some examples of which are given below.

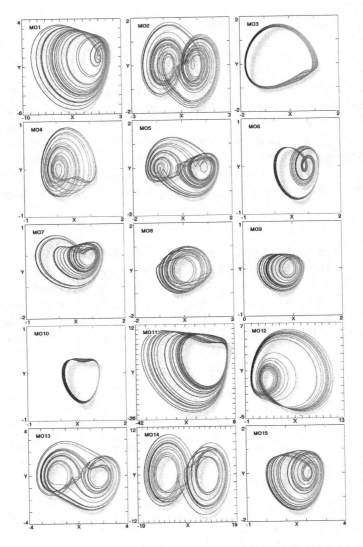

Fig. 3.21　Attractors for the chaotic memory oscillators in Table 3.3.

3.6.1　*Halvorsen's system*

A simple example of such a circulant system was suggested by Arne Dehli Halvorsen (unpublished) and given by

$$f(x, y, z) = -ax - 4y - 4z - y^2 \tag{3.22}$$

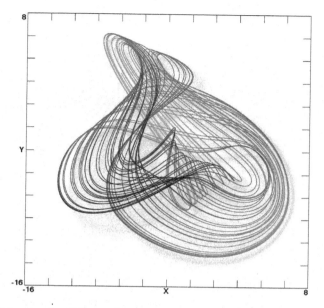

Fig. 3.22   Attractor for Halvorsen's circulant system from Eq. (3.22) for $a = 1.3$ and initial conditions $(x_0, y_0, z_0) = (-6.4, 0, 0)$, $\lambda = (0.6928, 0, -4.5928)$.

with a chaotic solution for $a = 1.3$ and an attractor as shown in Fig. 3.22 (Sprott, 2003). This system appears to be the simplest such example with a single quadratic nonlinearity.

3.6.2   *Thomas' systems*

René Thomas (1999) proposed a circulant case with one fewer term but with a cubic nonlinearity given by

$$f(x, y, z) = -ax + by - y^3, \tag{3.23}$$

which is chaotic for $a = 1$ and $b = 4$ with an attractor as shown in Fig. 3.23. This system has 27 equilibrium points, all unstable. It was inspired by another such system that Thomas proposed and given by

$$f(x, y, z) = -ax + \sin y, \tag{3.24}$$

which is chaotic over most of the range $a < 0.1$ as well as some other values such as $a = 0.18$ and has infinitely many equilibrium points. The first two terms of the expansion of $\sin y$ are given by $\sin y \approx y - y^3/6$, which leads to

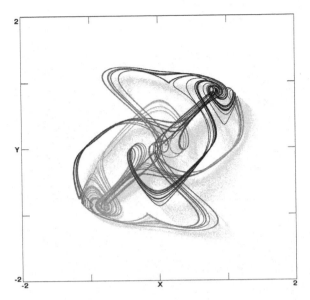

Fig. 3.23   Attractor for Thomas' circulant system from Eq. (3.23) for $a = 1$ and $b = 4$ with initial conditions $(x_0, y_0, z_0) = (0.4, 0, 0)$, $\lambda = (0.1785, 0, -3.1785)$.

a system of the form of Eq. (3.23). Discussion of Eq. (3.24) will be deferred to Chapter 4 because it has the unusual property of exhibiting chaos for $a = 0$, which is especially elegant but not dissipative.

3.6.3   *Piecewise-linear system*

A final example of a three-dimensional circulant system is the piecewise-linear system given by

$$f(x, y, z) = 1 - x - y - 4|y| \tag{3.25}$$

with an attractor shown in Fig. 3.24. This system also has a conservative variant that will be described in Chapter 4.

3.7   Other Systems

We conclude this chapter with a number of other three-dimensional chaotic systems that have been proposed and studied, some of which have potential applications.

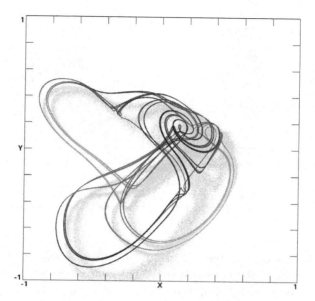

Fig. 3.24 Attractor for the piecewise-linear circulant system from Eq. (3.25) with initial conditions $(x_0, y_0, z_0) = (0.4, 0, 0)$, $\lambda = (0.0975, 0, -3.0975)$.

3.7.1 *Multiscroll systems*

The Lorenz attractor in Fig. 3.1 is an example of a *two-scroll* system in that the attractor has two lobes, like the wings of a butterfly. Other two-scroll systems include cases SQ_B and SQ_C in Fig. 3.9, the Moore–Spiegel system in Fig. 3.14, and cases MO_2, MO_5, MO_{13}, and MO_{14} in Fig. 3.21. The three-dimensional circulant systems just discussed have three lobes because of their three-fold symmetry. Apparently such systems require either multiple quadratic nonlinearities or a single nonlinearity of a higher order such as a cubic. It is interesting to ask if there are simpler three-scroll systems with quadratic nonlinearities that are not circulant.

Perhaps the simplest such example is

$$\begin{aligned}
\dot{x} &= x - yz \\
\dot{y} &= -y + xz \\
\dot{z} &= -3z + xy,
\end{aligned}$$

(3.26)

which resembles a circulant system except for the signs and the factor of 3 in the $\dot{z}$ equation. Its attractor as shown in Fig. 3.25(a) looks like a smiley face.

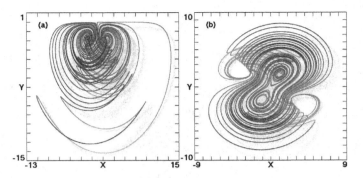

Fig. 3.25   (a) Attractor for the three-scroll system in Eq. (3.26) with $(x_0, y_0, z_0) = (1, -2, 0)$, $\lambda = (0.3779, 0, -3.3779)$ (b) Attractor for the four-scroll system in Eq. (3.27) with $(x_0, y_0, z_0) = (1, 0, 0)$, $\lambda = (0.2484, 0, -3.2484)$.

Three-dimensional systems with three quadratic nonlinearities can also produce *four-scroll* attractors (Wang, 2009), one of the simplest examples of which is obtained by simply adding a $+x$ term to the $\dot{y}$ equation in Eq. (3.26) to obtain

$$\dot{x} = x - yz$$
$$\dot{y} = x - y + xz \qquad\qquad (3.27)$$
$$\dot{z} = -3z + xy$$

whose attractor is shown in Fig. 3.25(b). Two of the lobes are a bit difficult to visualize because they lie out of the plane of the page. It is likely that systems with more than four lobes require a higher dimension or a more complicated nonlinearity such as a trigonometric function or a piecewise linear system with many breakpoints (Yalçin *et al.*, 2005; Lü and Chen, 2006).

3.7.2   *Lotka–Volterra systems*

Of particular interest to ecologists are population-dynamic models in which N species are interacting with one another through some combination of predation, competition, and cooperation. Such models were studied long ago by Lotka (1925) and independently by Volterra (1926), and are described by

$$\dot{x}_i = r_i x_i \left(1 - \sum_{j=1}^{N} a_{ij} x_j\right). \qquad\qquad (3.28)$$

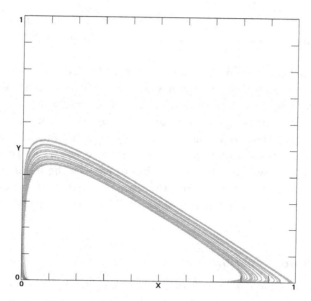

Fig. 3.26   Lotka–Volterra attractor from Eq. (3.29) with initial conditions $(x_0, y_0, z_0) = (0.6, 0.2, 0.01)$, $\lambda = (0.0122, 0, -0.3322)$.

While these equations are quadratic with a certain symmetry, they contain many nonlinearities and a generally complicated matrix of coefficients a_{ij} and a vector of growth rates r_i wherein reside all the biological detail. Although these equations have been criticized for ignoring many important effects, they can be viewed as the first nonlinear approximation in a Taylor-series expansion for a wide class of more realistic models (McArthur, 1970).

It has been shown (Arneodo *et al.*, 1980) that Eq. (3.28) can exhibit chaos for $N \geq 3$. One of the simplest examples is given by

$$\dot{x} = x(1 - x - 9y)$$
$$\dot{y} = -y(1 - 6x - y + 9z)$$
$$\dot{z} = z(1 - 3x - z)$$

(3.29)

whose attractor is shown in Fig. 3.26. The three species oscillate almost periodically but out of phase, with each species in turn nearly vanishing while the other two are abundant and in competition. Such behavior is not uncommon in nature because if a species becomes too rare, its predators cease searching for it. Furthermore, in a real ecology with a species under great stress, only the fittest individuals will survive. Note that if a species ever completely vanishes, it can never recover, and the dimension of the system decreases by one, which in this case would preclude the possibility of chaos.

Although this system is simple as Lotka–Volterra models go, and it illustrates that ecologies with as few as three species with only seven of the possible nine interactions (including self-interactions) and the same magnitudes of growth rates can exhibit chaos, it is not very simple by the standards of this book since it contains seven nonlinearities. Chaotic Lotka–Volterra models of this type also require a careful tuning of parameters and are even less likely to exhibit chaos as the number of species increases (Sprott *et al.*, 2005a), but there are some simple counter-examples (Sprott *et al.*, 2005b; Wildenberg *et al.*, 2005), one of which is discussed in Chapter 7.

3.7.3 *Chua's systems*

In the 1980s, Leon Chua and colleagues developed a class of electronic circuit capable of exhibiting chaos with a wide range of behaviors (Matsumoto *et al.*, 1985). The system has been extensively studied by many people (Chua, 1994), and there is even a book devoted to it containing a gallery of nearly 900 strange attractors (Bilotta and Pantano, 2008). The usual variant of the circuit is modelled by the equations

$$\dot{x} = \alpha[y - x + bx + 0.5(a - b)(|x + 1| - |x - 1|)]$$
$$\dot{y} = x - y + z \tag{3.30}$$
$$\dot{z} = -\beta y - \gamma z.$$

With nine terms and five parameters, it is not very mathematically elegant compared with other cases in this chapter.

However, a more elegant variant with unnecessary terms removed and others normalized to 1.0 is given by

$$\dot{x} = ay - x + |x + 1| - |x - 1|$$
$$\dot{y} = z - x \tag{3.31}$$
$$\dot{z} = y.$$

The system now contains a single parameter with chaos for $a = 3$. Its attractor is shown in Fig. 3.27.

A number of other variants of Chua's circuit have been proposed and studied, some of which are listed in Table 3.4. Khibnik *et al.* (1993), Zhong

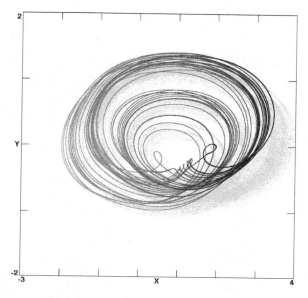

Fig. 3.27 Attractor for an elegant form of Chua's circuit in Eq. (3.31) with $a = 3$ and initial conditions $(x_0, y_0, z_0) = (1, 0, 0)$, $\lambda = (0.1131, 0, -0.3948)$.

Table 3.4 Variants of Chua's circuit with $\dot{y} = x + z$ and $\dot{z} = -y$.

Model	Equation	x_0, y_0, z_0	Lyapunov Exponents
CV_1	$\dot{x} = 0.3y + x - x^3$	$0, -3, 1$	$0.1269, 0, -0.9489$
CV_2	$\dot{x} = 0.2y - x + 2\tanh x$	$0, 1, -6$	$0.0556, 0, -0.1479$
CV_3	$\dot{x} = 0.2y + x - x|x|$	$0, 1, -3$	$0.0518, 0, -0.2233$
CV_4	$\dot{x} = 0.2y - x + 2\sin x$	$0, 6, 0$	$0.0504, 0, -0.5017$
CV_5	$\dot{x} = 0.2y - 0.3x + \text{sgn } x$	$0, 4, -8$	$0.0288, 0, -0.1540$
CV_6	$\dot{x} = 0.2y - x + 2\arctan x$	$0.7.6, 0$	$0.0095, 0, -0.0952$

(1994), Tsuneda (2005), and others have replaced the piecewise-linear function of x in the $\dot{x}$ equation with a cubic nonlinearity, which has a similar but smoother shape. Its attractor is shown in model CV_1 of Fig. 3.28. This system is relatively simple since it has only six terms and a single cubic nonlinearity. Alternately, a sigmoid function such as the hyperbolic tangent can be used for the nonlinearity (Brown, 1993; Özoğuz *et al.*, 2002) as shown in model CV_2 of Fig. 3.28. The nonlinear function can be replaced by $x|x|$ (Tang *et al.*, 2001a) as shown in model CV_3 of Fig. 3.28 or by a sine function (Tang *et al.*, 2001b) as shown in model CV_4 of Fig. 3.28. A discontinuous (signum) nonlinearity can be used (Mahla and

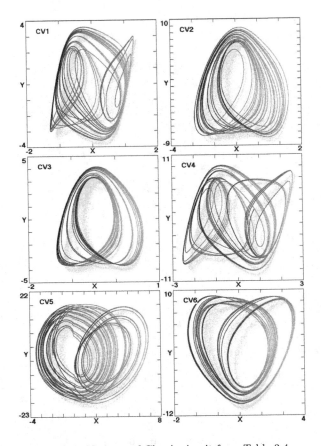

Fig. 3.28   Variants of Chua's circuit from Table 3.4.

Badan Palhares, 1993; Elawkil *et al.*, 2000) as shown in model CV_5 of Fig. 3.28. A variety of other nonlinear functions also suffice to produce chaos in Chua's system, as for example in the model CV_6 of Fig. 3.28, which involves an arc tangent.

3.7.4   *Rikitake dynamo*

The frequent and irregular reversals of the Earth's magnetic field motivated a number of early studies involving electrical currents within the Earth's molten core. One of the first such models to exhibit reversals was Rikitake's two-disk dynamo (Rikitake, 1958), which predated the work of Lorenz and is here given special attention because of its historical importance and

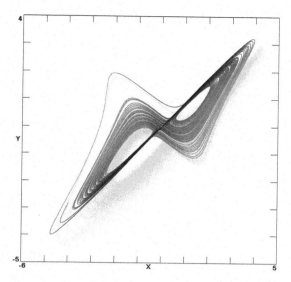

Fig. 3.29 Attractor for the Rikitake dynamo in Eq. (3.32) with $\mu = \alpha = 1$ and initial conditions $(x_0, y_0, z_0) = (1, 0, 0.6)$, $\lambda = (0.1273, 0, -2.1273)$.

potential application. An earlier single-disk dynamo model had been proposed by Bullard (1955), but it was incapable of producing the polarity reversals that are known to have occurred. Rikitake observed complex solutions of his model equations, but he could not follow the solutions for a sufficient time to establish aperiodicity, much less chaos. In a later study, Allan (1962) observed signs of aperiodicity, but it was not until Cook and Roberts (1970) carried out extensive numerical computations on the Rikitake system that chaos was firmly established and the similarity to the Lorenz system was noted. Much more extensive studies were subsequently carried out by Ito (1980) and by Hoshi and Kono (1988).

The equations that govern the Rikitake dynamo are

$$\dot{x} = -\mu x + yz$$
$$\dot{y} = -\mu y + x(z - \alpha) \qquad (3.32)$$
$$\dot{z} = 1 - xy,$$

which has the same number of terms as the Lorenz system, but with one additional nonlinearity. The parameter μ represents resistive dissipation, and the parameter α represents the difference in the angular velocities of the two disks. It turns out that both parameters can be set to 1.0 without destroying the chaos, leading to the two-scroll attractor shown in Fig. 3.29.

Chapter 4

Autonomous Conservative Systems

Systems without dissipation have a long history rooted in the study of celestial mechanics and formalized by Euler, Lagrange, Hamilton, Jacobi, and others two centuries ago. Although such systems do not have attractors, they can exhibit chaos, some examples of which were given in the forced oscillators in Chapters 1 and 2. Here we consider autonomous conservative systems with only three variables and describe some of the most elegant such systems that exhibit chaos.

4.1 Nosé–Hoover Oscillator

In principle, any of the autonomous dissipative systems in the previous chapter could exhibit chaos in the limit where the damping is reduced to zero, but in practice, they seldom do, unlike the periodically forced systems that often remain chaotic in the limit of zero damping. For example, the famous Lorenz system in Eq. (3.1) would be conservative if the coefficient of the x term in the $\dot{x}$ equation, the coefficient of the y term in the $\dot{y}$ equation, and the coefficient of the z term in the $\dot{z}$ equation summed to zero. No such systems are found to exhibit chaos for any choice of the other coefficients.

However, there are some conservative three-dimensional autonomous systems with quadratic nonlinearities that do exhibit chaos. Perhaps the most elegant such system is the Nosé–Hoover oscillator (Nosé, 1991; Hoover, 1995) given by

$$\dot{x} = y$$
$$\dot{y} = yz - x \qquad (4.1)$$
$$\dot{z} = 1 - y^2.$$

This system has only five terms and two quadratic nonlinearities. It

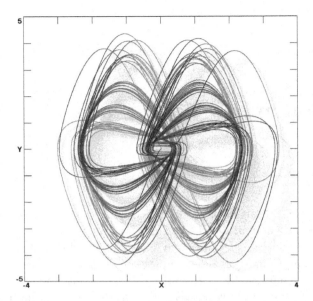

Fig. 4.1 Nosé–Hoover oscillator from Eq. (4.1) for $(x_0, y_0, z_0) = (0, 0.1, 0)$, $\lambda = (0.0139, 0, -0.0139)$.

was independently discovered in the systematic search for simple three-dimensional chaotic systems with quadratic nonlinearities (Sprott, 1994) that resulted in Table 3.1, and it can be considered as model SQ_A in that listing. It was the only such conservative chaotic system found with five terms and two quadratic nonlinearities or six terms and one quadratic nonlinearity, and hence it is likely to be the simplest such system. In fact, Heidel and Zhang (1999) have proved that there can be no simpler case.

This system is not obviously conservative since the dissipation is given by the time-averaged value of z along the trajectory, but numerical calculations indicate that this quantity is zero to a high precision. Furthermore, the system is invariant to the transformation $t \to -t$ (provided also $y \to -y$ and $z \to -z$), and hence it is time reversible with Lyapunov exponents that are symmetric about zero. The state space plot for Eq. (4.1) is shown in Fig. 4.1.

For such a conservative system in a three-dimensional state space, the chaotic trajectory wanders around in a three-dimensional sea with boundaries ('coastlines') outside of which the trajectories are typically quasiperiodic and lie on surfaces of nested toruses. Some initial conditions lie within the sea, while others lie on the toruses. For these reasons, it is more instruc-

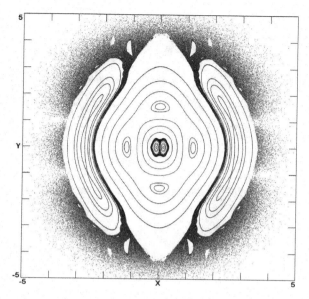

Fig. 4.2 Poincaré section for the Nosé–Hoover oscillator from Eq. (4.1) in the $z = 0$ plane.

tive to plot a Poincaré section of the trajectory, for example by showing those points in the xy-plane where the trajectory crosses the $z = 0$ plane, and to do it for many different initial conditions so as to capture both kinds of behavior. In such a section, the three-dimensional chaotic sea will be a two-dimensional region of the plane populated with what appear to be random dots, and the quasiperiodic trajectories will puncture the plane along closed curves called *drift rings*. Figure 4.2 shows such a plot for Eq. (4.1). Note the light bands in the vicinity of $y = \pm 1$ where the points are sparse resulting from the fact that $\dot{z} = 0$ at $y = \pm 1$, and hence the trajectories are tangent to the $z = 0$ plane there.

4.2 Nosé–Hoover Variants

A variant of the Nosé–Hoover system allows the y^2 term in the $\dot{z}$ equation to have a different exponent b so that the equations become

$$\dot{x} = y$$
$$\dot{y} = yz - x \qquad (4.2)$$
$$\dot{z} = 1 - |y|^b$$

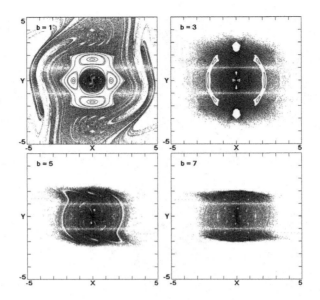

Fig. 4.3   Some Poincaré sections for the Nosé–Hoover variant from Eq. (4.2) in the $z = 0$ plane.

where b is not necessarily an integer. In fact, chaos apparently occurs for all $b \geq 1$, some Poincaré sections of which are shown in Fig. 4.3. Note that increasing b shrinks the quasiperiodic regions while keeping the trajectories in the chaotic sea closer to the origin.

4.3   Jerk Systems

Just as it was possible to reduce a wide variety of three-dimensional dissipative systems to jerk form, the same is true of conservative systems, and that represents in some sense a simplification and often makes the resulting equations more elegant.

4.3.1   *Jerk form of the Nosé–Hoover oscillator*

Gottlieb (1996) pointed out that the Nosé–Hoover oscillator can be written in jerk form as

$$\dddot{x} = \frac{(\ddot{x} + x)\ddot{x}}{\dot{x}} - \dot{x}^3, \tag{4.3}$$

which is a compact way of writing the three-dimensional system

$$\dot{x} = y$$
$$\dot{y} = z \tag{4.4}$$
$$\dot{z} = (z + x)z/y - y^3.$$

Although Eq. (4.4) is equivalent to Eq. (4.1) at least in its projection onto the xy-plane, it is hardly a simplification since it also contains five terms, but now with three nonlinearities, one of which is cubic and the other two involve $1/y$. The latter pose numerical difficulties since the trajectory repeatedly crosses the $y = 0$ plane where the value of $\dot{z}$ is infinite. Fortunately, it is not necessary to pursue this example since it is clearly less elegant than the equivalent polynomial form in Eq. (4.1)

4.3.2 *Simplest conservative chaotic flow*

The simplest conservative jerk system with chaotic solutions appears to be

$$\dddot{x} + 8\dot{x} = |x| - 1 \tag{4.5}$$

whose state space plot is shown in Fig. 4.4. This system only very slightly departs from a torus as evidenced by the appearance of the state space plot and by its small Lyapunov exponents. The Poincaré section for $z = 0$ in Fig. 4.5 shows that most initial conditions produce solutions that lie on the surface of nested toruses, but that there is a thin chaotic sea.

It is tempting to suppose that there might be chaotic systems of the form of Eq. (4.5) but with the $|x|$ term replaced by x^2. In fact, such cases have been studied (Sprott, 1997b; Heidel and Zhang, 2007), but the chaotic sea appears to stretch to infinity.

4.3.3 *Other conservative jerk systems*

The system in Eq. (4.5) is only the first in an hierarchy of chaotic conservative jerk systems, some other examples of which are listed in Table 4.1 (Sprott, 2003). The state space plots for these systems are shown in Fig. 4.6. Most of these systems have a rather small chaotic sea, more like a 'chaotic pond' and correspondingly small Lyapunov exponents. As a result, initial conditions must be chosen carefully to observe the chaos as indicated in the table, and even then, the trajectory tends to lie close to a torus as suggested by the figure.

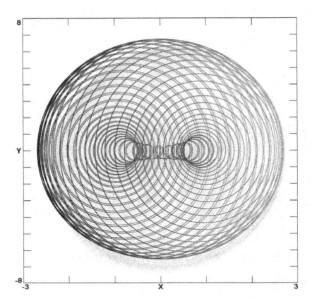

Fig. 4.4   Simplest chaotic conservative jerk system from Eq. (4.5) for $(x_0, y_0, z_0) = (0.38, 0.01, 1.62)$, $\lambda = (0.0022, 0, -0.0022)$.

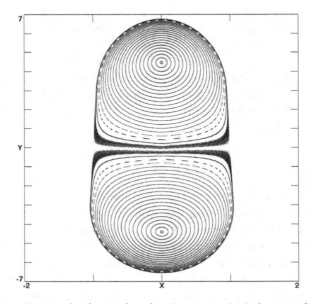

Fig. 4.5   Poincaré section for the simplest chaotic conservative jerk system from Eq. (4.5) in the $z = 0$ plane.

Table 4.1 Chaotic conservative jerk systems.

Model	Equation	$x_0, \dot{x}_0, \ddot{x}_0$	Lyapunov Exponents
CJ_0	$\dddot{x} + 8\dot{x} = |x| - 1$	$0.38, 0.01, 1.62$	$0.0022, 0, -0.0022$
CJ_1	$\dddot{x} + 4\dot{x} = x(x^2 - 1)$	$-0.4, 0.2, -0.8$	$0.0020, 0, -0.0020$
CJ_2	$\dddot{x} + \dot{x} = 0.5x - \max(x, 0) + 1$	$-0.58, -3.53, 0.61$	$0.0014, 0, -0.0014$
CJ_3	$\dddot{x} + 5\dot{x} = 2x - \sinh x$	$-1.22, -0.04, 1.43$	$0.0016, 0, -0.0016$
CJ_4	$\dddot{x} + 9\dot{x} = 3 - \cosh x$	$-0.61, -1.73, 1.65$	$0.0023, 0, -0.0023$

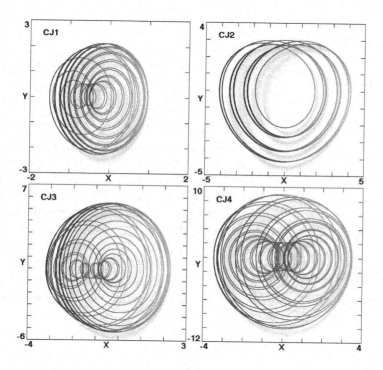

Fig. 4.6 State space plots for some of the conservative jerk systems in Table 4.1.

4.4 Circulant Systems

Just as Eq. (3.21) allows cyclically symmetric dissipative chaotic solutions in three-dimensional state space, it can also produce conservative chaotic solutions. One way to ensure no dissipation is to make each derivative depend only on the other two state space variables according to

$$\dot{x} = f(y, z)$$
$$\dot{y} = f(z, x) \qquad\qquad (4.6)$$
$$\dot{z} = f(x, y)$$

where the functions are all identical except the variables are rotated. Many nonlinear functions are capable of producing chaos for such a system, a few simple examples of which are given here.

4.4.1   Quadratic case

This simplest such case contains a single quadratic nonlinearity and is given by

$$f(y, z) = y^2 - z \qquad\qquad (4.7)$$

and its variants resulting from cycling the variables in the opposite direction ($f(y, z) = z^2 - y$) and changing their signs ($f(y, z) = z - y^2$). Its state space plot is shown in Fig. 4.7. The Poincaré section for $z = 0$ in Fig. 4.8 shows that most initial conditions produce solutions that lie on the surface of nested toruses, but there is a thin chaotic sea just barely visible near the outer edge of the region of confined orbits.

4.4.2   Cubic case

This simplest cubic circulant case that exhibits chaos is similar to the quadratic case and is given by

$$f(y, z) = y^3 - z \qquad\qquad (4.8)$$

and variants resulting from rotating the variables in the opposite direction and changing their signs. Its state space plot as shown in Fig. 4.9 lies near the surface of a torus. The Poincaré section for $z = 0$ in Fig. 4.10 resembles that in Fig. 4.8 except that the islands are somewhat wider in the vicinity of the chaotic sea, which is not evident in the figure.

Equations (4.7) and (4.8) suggest that there might be a family of chaotic systems of the form $f(y, z) = |y|^b - z$, where b is not necessarily an integer. Indeed, such appears to be the case for $b > 1$.

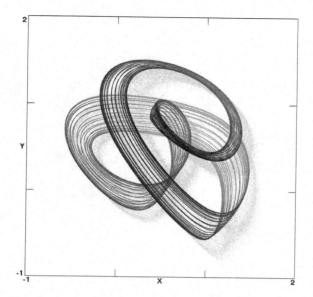

Fig. 4.7 Simplest chaotic quadratic conservative circulant system from Eq. (4.7) for $(x_0, y_0, z_0) = (0.7, 0.6, 1.2)$, $\lambda = (0.0070, 0, -0.0070)$.

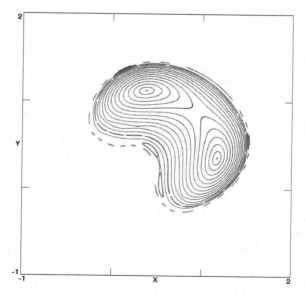

Fig. 4.8 Poincaré section for the simplest chaotic quadratic conservative circulant system from Eq. (4.7) in the $z = 0$ plane.

Elegant Chaos

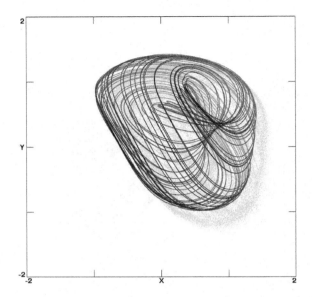

Fig. 4.9 Simplest chaotic cubic conservative circulant system from Eq. (4.8) for $(x_0, y_0, z_0) = (-0.77, 0.35, 1.13)$, $\lambda = (0.0059, 0, -0.0059)$.

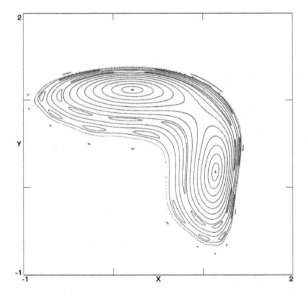

Fig. 4.10 Poincaré section for the simplest chaotic cubic conservative circulant system from Eq. (4.8) in the $z = 0$ plane.

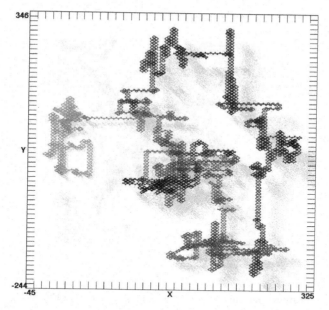

Fig. 4.11   Labyrinth chaos showing fractional Brownian motion from Eq. (4.9) for $(x_0, y_0, z_0) = (1, 0, 0)$, $\lambda = (0.0921, 0, -0.0921)$.

4.4.3   *Labyrinth chaos*

René Thomas (1999) proposed a particularly elegant chaotic system given by

$$\dot{x} = \sin y$$
$$\dot{y} = \sin z \qquad\qquad (4.9)$$
$$\dot{z} = \sin x,$$

which he called 'labyrinth chaos' because the state space is divided into three-dimensional cells by the sinusoidal nonlinearities, and the trajectory wanders throughout the state space without bound in a kind of pseudo-random walk as shown in Fig. 4.11. Equation (4.9) is the conservative ($a = 0$) limit of the case in Eq. (3.24). Its Poincaré section is shown in Fig. 4.12.

This case has been studied in some detail by Sprott and Chlouverakis (2007) who find that the chaotic sea occupies 98.33% of the state space and that the diffusion obeys fractional Brownian motion, enhanced by the intermittency that occurs when the trajectory comes close to one of a number of small quasiperiodic 'worm holes' that occupy the remainder of the space.

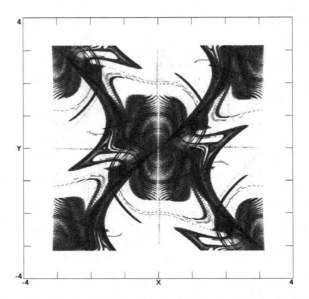

Fig. 4.12   Poincaré section (mod 2π) for the chaotic labyrinth in Eq. (4.9) in the z (mod 2π) $= 0$ plane.

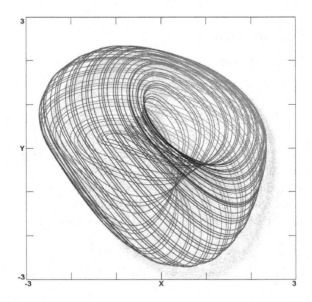

Fig. 4.13   Piecewise-linear conservative circulant system from Eq. (4.10) with initial conditions $(x_0, y_0, z_0) = (0.3, -0.9, -1.4)$, $\lambda = (0.0022, 0, -0.0022)$.

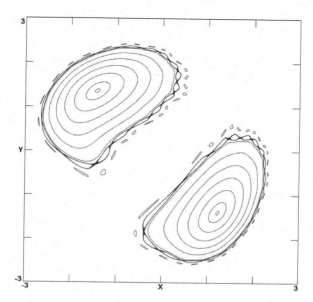

Fig. 4.14 Poincaré section for the piecewise-linear conservative circulant system from Eq. (4.10) in the $z = 0$ plane.

Rowlands and Sprott (2008) further examined the anomalous diffusion scaling that occurs in this system and were able to quantify the departure from a simple random walk.

4.4.4 *Piecewise-linear system*

We conclude with a final example of a three-dimensional circulant piecewise-linear system given by

$$f(x, y, z) = 1 - y + z - |y|, \qquad (4.10)$$

which is a conservative variant of Eq. (3.25) whose state space plot is shown in Fig. 4.13 and Poincaré section is shown in Fig. 4.14. The chaotic sea is localized to a very thin region in the vicinity of the separatrix in Fig. 4.14, and thus the initial conditions for chaotic solutions are rather critical.

Chapter 5

Low-dimensional Systems ($D < 3$)

This brief chapter will describe some systems in two-dimensional state space that exhibit a kind of behavior that could be described as chaotic in apparent violation of the Poincaré–Bendixson theorem. These systems have in their flow a singularity that acts as a scattering center and makes the dynamics sensitive to an arbitrarily small amount of noise, either experimental or numerical. Such systems have been relatively little studied and are thus ripe for further investigation.

5.1   Dixon System

The systems in this chapter were inspired by the work of Dixon *et al.* (1993) in which they transformed a set of three ODEs introduced by Cummings *et al.* (1992) to model the dynamical behavior of the magnetic field of a neutron star. The transformation reduced the system to a two-dimensional flow that nevertheless preserves the chaotic behavior in apparent violation of the Poincaré–Bendixson theorem (Hirsch *et al.*, 2004), which states that the attractor for any smooth two-dimensional bounded continuous-time autonomous system is either a stable equilibrium or a limit cycle.

The system derived by Dixon *et al.* (1993) is given by

$$\dot{x} = \frac{xy}{x^2 + y^2} - \alpha x$$

$$\dot{y} = \frac{y^2}{x^2 + y^2} - \beta y + \beta - 1 \tag{5.1}$$

and is singular at the origin ($x = y = 0$) and thus does not satisfy the smoothness condition required for the Poincaré–Bendixson theorem to apply. All orbits are attracted to the singularity in finite time, and as a result

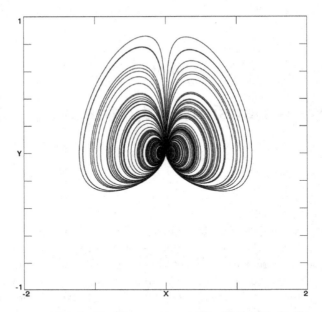

Fig. 5.1   State space plot for the Dixon system in Eq. (5.1) with $(\alpha, \beta) = (0, 0.7)$ for $(x_0, y_0) = (1, 0)$.

they are sensitive to even the smallest nonzero perturbation, including one arising from the numerical method. The system displays what Dixon *et al.* (1993) call 'S-chaos' (singularity-chaos) for a range of parameters including $\alpha = 0$ and $\beta = 0.7$. The state space plot for this case is shown in Fig. 5.1. A longer calculation of the trajectory would show that it densely fills a region of the xy-plane, and hence it is two-dimensional. Since the resulting plot does not have fractal structure, it is not a proper strange attractor, and the dynamics perhaps should not be considered truly chaotic as pointed out by Alvarez-Ramirez, *et al.* (2005). For the same reason, Lyapunov exponents are not quoted in this chapter because they are not well-defined, and their calculation in the vicinity of the singularity is problematic.

5.2   Dixon Variants

Without straying too far from the form of the Dixon system in Eq. (5.1), one can construct a number of slightly simplified variants, some of which are given in Table 5.1 along with model DV_0 which was shown in Fig. 5.1.

Table 5.1   Simplified variants of the Dixon system.

Model	Equations		x_0, y_0
DV_0	$\dot{x} = \dfrac{xy}{x^2 + y^2}$	$\dot{y} = \dfrac{y^2}{x^2 + y^2} - 0.7y - 0.3$	$0.1, 0$
DV_1	$\dot{x} = \dfrac{xy}{x^2 + y^2}$	$\dot{y} = \dfrac{y^2}{x^2 + y^2} - y - 0.3$	$0.1, 0$
DV_2	$\dot{x} = \dfrac{xy}{x^2 + y^2}$	$\dot{y} = \dfrac{y^2}{x^2 + 0.6y^2} - y - 1$	$0.1, 0$
DV_3	$\dot{x} = \dfrac{xy}{x^2 + y^2} - x$	$\dot{y} = \dfrac{y^2}{x^2 + y^2} - y - 0.1$	$0.1, 0$
DV_4	$\dot{x} = \dfrac{xy}{2x^2 + y^2}$	$\dot{y} = \dfrac{y^2}{x^2 + y^2} - 0.5$	$0.1, 0$

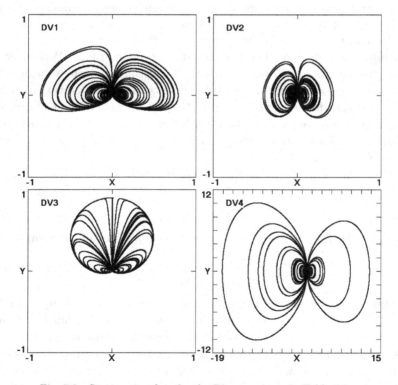

Fig. 5.2   State space plots for the Dixon variants in Table 5.1.

State space plots for the cases in Table 5.1 as shown in Fig. 5.2 display an apparent similarity to the original Dixon system in Fig. 5.1.

5.3   Logarithmic Case

A weaker singularity is $\log |x|$, which has a value of $-\infty$ at $x = 0$. A simple two-dimensional example that appears to exhibit S-chaos is given by

$$\dot{x} = \log |x| - y$$
$$\dot{y} = \log |x| + x. \tag{5.2}$$

Although it is not critical, the logarithm is here taken as base-e. Note that the absolute value $|x|$ is required to avoid taking the logarithm of a negative number whose value would not be real. The absolute value can be removed by using the relation $\log |x| = 0.5 \log x^2$.

In the absence of the singularity, this system is just a harmonic oscillator with the orbit tracing a counter-clockwise circle in the xy-plane. However, the logarithmic term is positive over most of the orbit (whenever $|x| > 0$), causing a slow drift of the circle toward the upper right. Whenever the orbit crosses the $x = 0$ axis where the logarithm is strongly negative, it gets a compensating kick downward and to the left, accelerating it slightly for $y > 0$ and decelerating it slightly for $y < 0$. The amount of acceleration/deceleration depends on how much time is spent in the vicinity of the singularity, which varies slightly with each orbit. The result is the state space plot shown in Fig. 5.3.

Note that Eq. (5.2) can be combined into a single second-order autonomous ODE of the form

$$\ddot{x} = \dot{x}/x - \log |x| + x, \tag{5.3}$$

which can then be written as two first-order equations of the form

$$\dot{x} = v$$
$$\dot{v} = v/x - \log |x| - x. \tag{5.4}$$

Because of the factor $1/x$, this system has a stronger singularity than does Eq. (5.2), and thus it is more difficult to solve numerically, but its behavior as shown in Fig. 5.4 is similar to Fig. 5.3, except the discontinuity at $x = 0$ is stronger and broader. In this case the orbit is traversed in the clockwise direction, and the downward acceleration from the v/x term when the orbit approaches the $x = 0$ axis is cancelled by the upward acceleration when it departs the $x = 0$ axis, leaving only the contribution from the $\log |x|$ term

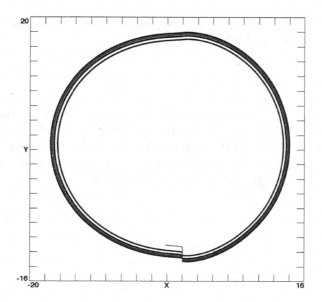

Fig. 5.3 State space plot for the logarithmic case in Eq. (5.2) with $(x_0, y_0) = (11, 0)$.

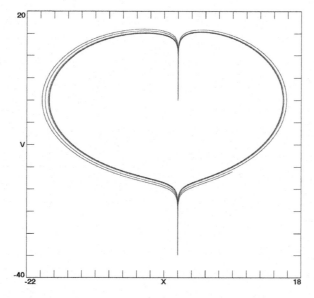

Fig. 5.4 State space plot for the logarithmic case in Eq. (5.4) with $(x_0, y_0) = (11, \log(11))$.

that causes a brief upward acceleration for $|x| < 1$ and offsets the downward drift that it produces for $|x| > 1$.

5.4 Other Cases

This chapter was originally intended to be considerably longer, but a search for additional chaotic two-dimensional systems with singularities of various types at the origin failed to produce additional cases. It remains an unfinished task to identify the conditions under which chaos occurs in systems of this type and to find additional such examples.

Chapter 6

High-dimensional Systems ($D > 3$)

While chaos in continuous-time systems with smooth flows requires only three state space dimensions, there are elegant examples in higher-dimensional spaces, many of which are important because they better represent phenomena in the real world, which typically involve many interacting variables. Such systems can be constructed by adding variables to known three-dimensional chaotic systems, by coupling multiple low-dimensional systems, or by simply searching the larger parameter space of such systems for chaotic solutions. The challenge is to increase the dimension while retaining the elegance and to display the high-dimensional dynamics in a meaningful way.

6.1   Periodically Forced Systems

It was previously noted that any nonautonomous system can be made autonomous by adding a single state space variable.   In that way, two-dimensional periodically forced systems can be expressed as three-dimensional autonomous systems, many of which produce chaos if they are nonlinear. For example, the $A \sin \Omega t$ forcing function in an equation can be replaced with $A \sin z$ and $dz/dt = \Omega$. However, such a system will still contain a $\sin z$ nonlinearity. An alternate approach is to define a new quantity $y = A \sin \Omega t$ and generate $y(t)$ using *two* additional linearly coupled equations

$$\dot{y} = u$$
$$\dot{u} = -\Omega^2 y \qquad (6.1)$$

whose solution for initial conditions $(y_0, u_0) = (0, A\Omega)$ is $y = A \sin \Omega t$. Equation (6.1) is just an undamped harmonic oscillator, which is known to have sinusoidal solutions of the desired form.

Note that the parameter A disappears from the original system and becomes part of an initial condition in the new system. Furthermore, if the original forcing term had contained a phase factor $A\sin(\Omega t + \phi)$, that could also have been included in the initial conditions for the new system by using $(y_0, u_0) = (A\sin\phi, A\Omega\cos\phi)$. It is often the case that systems can be transformed in such a way that the parameters and initial conditions are interchanged. Alternately, think of the initial conditions as being additional parameters.

6.1.1 *Forced pendulum*

Perhaps the simplest such example comes from transforming the frictionless forced pendulum in Eq. (1.6) into the system

$$\begin{aligned} \ddot{x} + \sin x &= y \\ \ddot{y} &= -y, \end{aligned} \qquad (6.2)$$

which is a compact way of writing the four-dimensional system

$$\begin{aligned} \dot{x} &= v \\ \dot{v} &= y - \sin x \\ \dot{y} &= u \\ \dot{u} &= -y. \end{aligned} \qquad (6.3)$$

The new system has a single $\sin x$ nonlinearity, which emphasizes that the $\sin t$ nonlinearity in the periodically forced pendulum is not responsible for the chaos and by itself is insufficient to produce chaos. When the state space trajectory is plotted on a torus, the $\sin t$ term is simply the azimuthal angle around the torus.

Note that the last two equations in Eq. (6.3) constitute an independent subsystem that is not influenced by the first two equations. Such a case is called a *master–slave* system since the linear oscillator dictates the behavior of the nonlinear one, but not the reverse. It is said to be *unidirectionally coupled*. Think of the linear system as being a massive flywheel rotating with a constant angular velocity that can provide unlimited energy to the relatively small pendulum without being measurably affected by it.

The state space plot for Eq. (6.3) is shown in Fig. 6.1. The plot shows the trajectory in the xv-plane with the y variable indicated in shades of gray and by the shadow below and to the right. The u variable is ignored in the plot, which means that the plot is a projection of the four-dimensional dynamic onto the three-dimensional xvy-space.

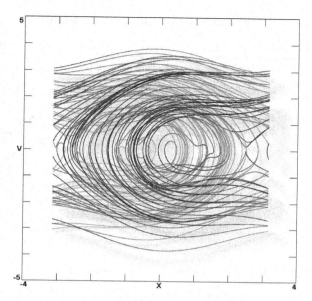

Fig. 6.1   State space plot for the conservative driven pendulum in Eq. (6.3) with initial conditions $(x_0, v_0, y_0, u_0) = (1, 0, 0, 1)$, $\lambda = (0.1502, 0, 0, -0.1502)$.

For many of the cases with four or more variables, only the largest Lyapunov exponent will be shown since the trick of calculating the spectrum from the largest exponent and their sum using the fact that one exponent must be zero no longer works when there are four exponents. However, for conservative cases such as this in four dimensions, one can use the fact that the Lyapunov exponents sum to zero and occur in equal and opposite pairs to calculate all four exponents from the largest and their sum. Since there will always be one zero exponent, corresponding to the direction of the flow, symmetry requires that there is a second zero exponent corresponding to a direction perpendicular to the flow in which there is neither contraction nor expansion of the state space. These two directions will typically represent a surface of constant total energy, which is nested with other such surfaces representing different values of the energy as determined by the initial conditions. Of course the entire spectrum of Lyapunov exponents can always be calculated using, for example, the method of Wolf *et al.* (1985), and that has been done for cases in this chapter where values for more than two nonzero exponents are quoted.

Table 6.1   Forced nonlinear oscillators with $\dot{y} = u$ and $\dot{u} = -\Omega^2 y$.

Model	Equations	x_0, v_0, y_0, u_0	Largest LE
FO_0	$\dot{x} = v, \dot{v} = -\sin x + y, \Omega = 1$	$1, 0, 0, 1$	0.1502
FO_1	$\dot{x} = v, \dot{v} = (1 - x^2)v - x + y, \Omega = 0.45$	$1, 1, 0, 0.45$	0.0389
FO_2	$\dot{x} = v, \dot{v} = 4v \cos x - 2x + y, \Omega = 0.8$	$-0.9, 4, 0, 4$	0.0837
FO_3	$\dot{x} = v, \dot{v} = (4 - v^2)v - x + y, \Omega = 4$	$1, 0.8, 0, 20$	0.1552
FO_4	$\dot{x} = v, \dot{v} = -v + (1 - x^2)x + y, \Omega = 0.8$	$-1, -0.6, 0, 0.8$	0.1216
FO_5	$\dot{x} = v, \dot{v} = -\operatorname{sgn} x + y, \Omega = 1$	$1, 0.1, 0, 1$	0.0462
FO_6	$\dot{x} = v, \dot{v} = -v + 1 - e^x + y, \Omega = 1$	$-24, -9, 0, 20$	0.0181
FO_7	$\dot{x} = v, \dot{v} = -x^3 + y, \Omega = 2$	$0, 0, 0, 2$	0.0905
FO_8	$\dot{x} = v + y, \dot{v} = -\sin x, \Omega = 1$	$0, 1, 0, 1$	0.1489
FO_9	$\dot{x} = v + y, \dot{v} = (1 - x^2)x, \Omega = 1$	$0, 1, 0, 1$	0.1152
FO_{10}	$\dot{x} = v, \dot{v} = y \sin x, \Omega = 1$	$0, 1, 0, 1$	0.1716
FO_{11}	$\dot{x} = v, \dot{v} = vy - x^3, \Omega = 1$	$1, 0, 0, 1$	0.0640
FO_{12}	$\dot{z} = \overline{z} + 1 - z^2 + u/\Omega + iy, \Omega = 1$	$0, -0.8, 0, 1$	0.0472

6.1.2   *Other forced nonlinear oscillators*

The method described above can be used for any periodically forced system such as those in Chapter 2, including velocity forced systems, parametrically forced systems and complex systems. A selection of such cases is given in Table 6.1 with their corresponding state space plots in Fig. 6.2. The plots are repetitive of those shown in Chapter 2, but the Lyapunov exponents are from independent calculations, and so any differences from those quoted in Chapter 2 are indicative of the precision of the numerics, but their similarity confirms the correctness of the transformation.

For the complex oscillator in FO_{12}, which except for a sign change is the same as model FZ_1 in Table 2.5, the complex variable z is assumed to be given by $z = x + iv$, and the forcing function is given by $e^{i\Omega t} = u/\Omega + iy$.

6.2   Master–slave Oscillators

There is no reason that the master oscillator has to be linear and the slave nonlinear. Of course to obtain chaos, at least one of the oscillators has to be nonlinear. However, when the linear oscillator is the slave, the nonlinear master oscillator cannot be chaotic if it is only two-dimensional, and so neither can the resulting system be chaotic since it consists of a periodically forced linear oscillator whose long-term solutions are necessarily periodic.

More interesting are cases in which both the master and slave are nonlinear. With dozens of examples of nonlinear oscillators in Chapter 2, there are

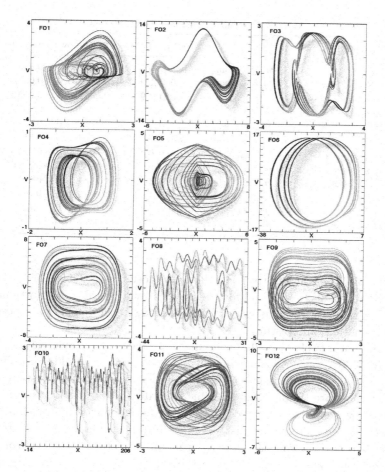

Fig. 6.2 State space plots for some of the forced nonlinear oscillators in Table 6.1.

hundreds of combinations of master–slaves, many of which have been studied in the literature (Chirikov, 1979; Gilmore and Letellier, 2007). Since we are interested here in elegant systems, only four of the simplest oscillators will be considered,

$$\ddot{x} + \sin x = F \tag{6.4}$$

$$\ddot{x} + \operatorname{sgn} x = F \tag{6.5}$$

$$\ddot{x} + x^3 = F \tag{6.6}$$

$$\ddot{x} = F \sin x, \tag{6.7}$$

Table 6.2   Chaotic master–slave oscillators.

Model	Equations	x_0, v_0, y_0, u_0	Lyapunov Exponents
MS_1	$\ddot{x} = -\sin x + y, \ddot{y} = -\sin y$	$0, 0, 0, 1.9$	$0.1761, 0, 0, -0.1767$
MS_2	$\ddot{x} = -\operatorname{sgn} x + y, \ddot{y} = -\sin y$	$0, 0, 0, 1.9$	$0.0058, 0, 0, -0.0058$
MS_3	$\ddot{x} = -x^3 + y, \ddot{y} = -\sin y$	$0, 0, 0, 1.9$	$0.0552, 0, 0, -0.0552$
MS_4	$\ddot{x} = y\sin x, \ddot{y} = -\sin y$	$1, 1, 0, 1.9$	$0.1914, 0, 0, -0.1914$
MS_5	$\ddot{x} = -\sin x + y, \ddot{y} = -\operatorname{sgn} y$	$0, 0, 0, 2$	$0.1728, 0, 0, -0.1677$
MS_6	$\ddot{x} = -\operatorname{sgn} x + y, \ddot{y} = -\operatorname{sgn} y$	$0, 1, 0.5, 0.75$	$0.0504, 0, 0, -0.0504$
MS_7	$\ddot{x} - x^3 + y, \ddot{y} = -\operatorname{sgn} y$	$0, 1, 0, 9$	$0.0455, 0, 0, -0.0455$
MS_8	$\ddot{x} = y\sin x, \ddot{y} = -\operatorname{sgn} y$	$1, 0, 0, 9$	$0.1090, 0, 0, -0.1090$
MS_9	$\ddot{x} = -\sin x + y, \ddot{y} = -y^3$	$0, 0, 0, 1$	$0.1581, 0, 0, -0.1581$
MS_{10}	$\ddot{x} = -\operatorname{sgn} x + y, \ddot{y} = -y^3$	$0, 0, 0, 1$	$0.0172, 0, 0, -0.0172$
MS_{11}	$\ddot{x} = -x^3 + y, \ddot{y} = -y^3$	$1, 0, 0, 1$	$0.0300, 0, 0, -0.0300$
MS_{12}	$\ddot{x} = y\sin x, \ddot{y} = -y^3$	$1, 0, 0, 1$	$0.1796, 0, 0, -0.1796$

all of which are conservative. For the master oscillator, the forcing function will be taken as $F = 0$, and for the slave, F will be taken proportional to the x variable of the master oscillator, although other choices are possible and potentially interesting.

There are twelve possible combinations since the parametric oscillator in Eq. (6.7) cannot be a master when $F = 0$ since it would not oscillate, and it would be equivalent to Eq. (6.4) if F were a nonzero constant. The twelve cases are summarized in Table 6.2 with their respective state space plots shown in Fig. 6.3. Note that since the master oscillator cannot be chaotic, these systems are just periodically forced nonlinear oscillators in which the forcing function has a harmonic spectrum resulting from the nonlinearity. For the cases that are driven by a pendulum ($\ddot{y} = -\sin y$), the initial conditions ($y_0 = 0, u_0 = 1.9$) are chosen so that the pendulum does not quite go 'over the top' (which would require $u_0 = 2$ with $y_0 = 0$) so as to enhance the harmonic content of F.

6.3   Mutually Coupled Nonlinear Oscillators

The previous examples are only apparently high-dimensional since two of the variables are not influenced by the others, and those examples are also somewhat nonphysical since they require one oscillator to have infinitely more energy than the other. More realistic cases involve two mutually coupled oscillators with similar energies.

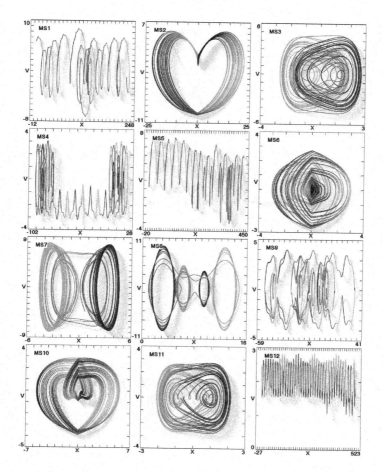

Fig. 6.3 State space plots for the master–slave oscillators in Table 6.2.

6.3.1 *Coupled pendulums*

A simple example is two identical frictionless pendulums coupled together by a torsional spring as shown in Fig. 6.4. The equations that govern such a system are

$$\ddot{x} + \sin x = k(y - x)$$
$$\ddot{y} + \sin y = k(x - y) \tag{6.8}$$

where k is the coupling constant, which is proportional to the stiffness of the spring. Although this is not the simplest form of coupling, it is the

Fig. 6.4   Two pendulums coupled together by a torsional spring.

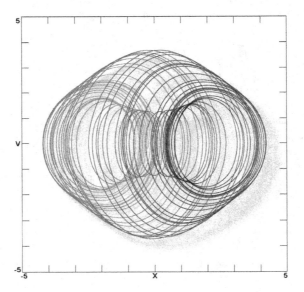

Fig. 6.5   State space plot for two identical frictionless coupled pendulums in Eq. (6.8) with $k = 1$ and initial conditions $(x_0, v_0, y_0, u_0) = (3.9, 0.3, 0.5, 0.3)$, $\lambda = (0.0057, 0, 0, -0.0057)$.

most physical since it satisfies Newton's third law, which states that the forces on two isolated objects interacting only with one another must be equal and opposite so as to conserve the total momentum of the system. The state space plot for this system with $k = 1$ is shown in Fig. 6.5.

It might seem logical to suppose that two identical coupled oscillators would simply lock together and oscillate with their displacements and velocities equal, and that often happens. However, it is sometimes the case that the synchronized state is unstable and the resulting behavior is chaotic when the oscillators are nonlinear as in the case here. Of course one has to be careful not to use identical initial conditions for the two oscillators since a perturbation to the unstable synchronized state is required for the instability to grow.

6.3.2 *Coupled van der Pol oscillators*

The previous example of coupled frictionless pendulums involves a conservative system. Dissipative systems can also be coupled in a similar manner. A widely studied such example (Rand and Holmes, 1980; Storti and Rand, 1982; Pastor *et al.*, 1993; Pastor-Díaz and Lopez-Fraguas, 1995; Low *et al.*, 2003; Camacho *et al.*, 2004) is the case of identical coupled van der Pol oscillators that individually produce limit cycles, but that when coupled together can have chaotic solutions much as does the periodically forced van der Pol system in Eq. (2.2).

A slightly more complicated coupling is required to produce chaos in such a case as given by

$$\ddot{x} + b(x^2 - 1)\dot{x} + x = k(y - \dot{y})$$
$$\ddot{y} + b(y^2 - 1)\dot{y} + y = k(x - \dot{x}), \tag{6.9}$$

which is chaotic for $b = 2.2$ and $k = 0.7$. The attractor for this case is shown in Fig. 6.6.

6.3.3 *Coupled FitzHugh–Nagumo oscillators*

Another limit-cycle oscillator was suggested by FitzHugh (1960, 1961) as a two-dimensional simplification of the four-dimensional *Hodgkin–Huxley model* of spike generation in squid giant axons (Hodgkin and Huxley, 1952). It was subsequently modelled electronically by Nagumo *et al.* (1962) and is now usually known as the *FitzHugh–Nagumo model*. Such a two-dimensional system cannot produce chaos, but two such systems coupled together can, one example of which is given by

$$\dot{x}_1 = x_1 - x_1^3 - y_1 + kx_2$$
$$\dot{y}_1 = a + bx_1 - cy_1$$
$$\dot{x}_2 = x_2 - x_2^3 - y_1 + kx_1$$
$$\dot{y}_2 = a + bx_2 - cy_2. \tag{6.10}$$

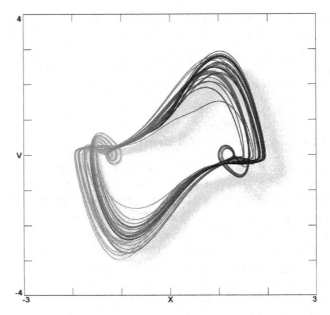

Fig. 6.6 Attractor for two identical coupled van der Pol oscillators in Eq. (6.9) with $b = 2.2, k = 0.7$, and initial conditions $(x_0, v_0, y_0, u_0) = (0, 1, 0, 1.7)$, $\lambda = (0.0829, 0, -0.3171, -2.1063)$.

In the limit of $k = a = b = c = 0$, the system reduces to two decoupled van der Pol oscillators (van der Pol, 1920), and thus FitzHugh (1961) originally called his system the *Bonhoeffer–van der Pol model* (Bonhoeffer, 1953), and it still sometimes goes by that name. Note that a simpler coupling suffices for this system than was required for the two coupled van der Pol oscillators in the previous section. The attractor for this case is shown in Fig. 6.7. In searching for elegant parameters, a constraint of $c > 0$ was imposed to keep the system from collapsing to a system of coupled van der Pol oscillators, although chaotic solutions require that c be rather small.

6.3.4 *Coupled complex oscillators*

Any of the nonlinear complex oscillators in Table 2.5 are candidates for coupling in a similar fashion to produce chaos. With seven cases from which to choose, there are $7 \times 7 = 49$ combinations of two systems, each of

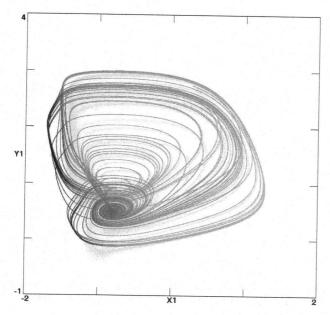

Fig. 6.7   Attractor for two identical coupled FitzHugh–Nagumo oscillators in Eq. (6.10) with $a = 0.4, b = 0.5, c = 0.01, k = -1$, and initial conditions $(x_{10}, y_{10}, x_{20}, y_{20}) = (-1.5, 1, 1, 1)$, $\lambda = (0.0281, 0, -0.2865, -2.4372)$.

which can be coupled in many different ways. For example, the dissipative system

$$\dot{z}_1 - z_1^2 - \overline{z}_1 + 1 = kz_2$$
$$\dot{z}_2 - z_2^2 - \overline{z}_2 + 1 = kz_1, \tag{6.11}$$

which resembles model FZ_1 in Table 2.5 (except for a change in sign of the z^2 terms) is chaotic for $k = -2.5$ with an attractor as shown in Fig. 6.8.

6.3.5   *Other coupled nonlinear oscillators*

Any of the other acceleration-forced systems in Chapter 2 can be mutually coupled in a manner similar to the previous cases. Table 6.3 gives a few such additional examples, all of which are conservative with their corresponding state space plots shown in Fig. 6.9. Model CO_0 has already been shown in Fig. 6.5.

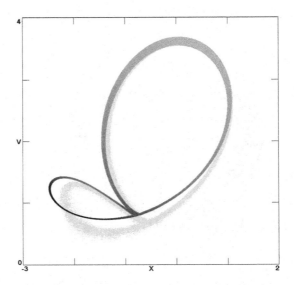

Fig. 6.8 Attractor for the coupled complex oscillators in Eq. (6.11) with $k = -2.5$ and initial conditions $(z_{10}, z_{20}) = (-2.4 + i, 1 - 3i)$, $\lambda = (0.0801, 0, -2.3684, -4.7164)$.

Table 6.3 Chaotic coupled nonlinear oscillators.

Model	Equations	x_0, v_0, y_0, u_0	Lyap Exponents
CO_0	$\ddot{x} + \sin x = y - x, \ddot{y} + \sin y = x - y$	$3.9, 0.3, 0.5, 0.3$	$0.0045, 0, 0, -0.0045$
CO_1	$\ddot{x} + \sin x = y - x, \ddot{y} + \operatorname{sgn} y = x - y$	$-0.2, 0.4, -2, -0.3$	$0.0177, 0, 0, -0.0177$
CO_2	$\ddot{x} + \sin x = y - x, \ddot{y} + y^3 = x - y$	$1.2, 1.9, 0.4, -1.2$	$0.0638, 0, 0, -0.0638$
CO_3	$\ddot{x} + \operatorname{sgn} x = y - x, \ddot{y} + \operatorname{sgn} y = x - y$	$0.8, 0.8, -2.1, 1.6$	$0.0642, 0, 0, -0.0642$
CO_4	$\ddot{x} + \operatorname{sgn} x = y - x, \ddot{y} + y^3 = x - y$	$-0.4, 0.5, -0.7, 1.3$	$0.0838, 0, 0, -0.0838$
CO_5	$\ddot{x} + x^3 = y - x, \ddot{y} + y^3 = x - y$	$1.7, 1.1, -3.2, 1.1$	$0.1468, 0, 0, -0.1468$

6.4 Hamiltonian Systems

Many conservative systems including those in Table 6.3 can be derived from Hamilton's equations (Arnold, 1978; Lichtenberg and Lieberman, 1992)

$$\dot{x} = \frac{\partial H}{\partial v}$$
$$\dot{v} = -\frac{\partial H}{\partial x} \tag{6.12}$$

(and similar equations for $(\dot{y}, \dot{u})$, and so forth), where the *Hamiltonian function* $H = H(x, v, y, u, \ldots)$ does not depend explicitly on time and is

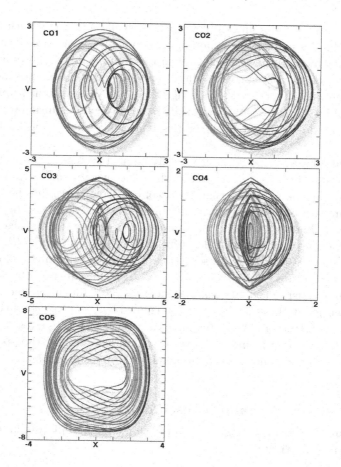

Fig. 6.9   State space plots for the coupled nonlinear oscillators in Table 6.3.

thus a constant of the motion, usually proportional to the total energy (kinetic plus potential) in the case of a mechanical system.

The Hamiltonian formulation of classical mechanics was introduced in 1833 by William Rowland Hamilton and parallels an earlier Lagrangian formulation introduced by Joseph Louis Lagrange in 1788. In fact, Hamilton's equations can be derived from Lagrange's equations. The Hamiltonian method expresses the dynamics as a system of first-order differential equations in a $2n$-dimensional phase space, whereas the Lagrangian method uses second-order equations in an n-dimensional coordinate space. Consequently, there is always an even number of Hamilton's equations, with the variables occurring in conjugate pairs, usually associated with the spatial coordinates and the corresponding components of the momentum. The equations have a natural elegance whose symmetry is broken only by the occurrence of a minus sign. The Hamiltonian description extends more naturally to quantum mechanics, although the position and momentum cannot be determined simultaneously to arbitrary accuracy in quantum mechanics, and the product of their uncertainties is a small value given by Planck's constant.

Hamilton's equations constitute a dynamical system in the usual sense used throughout this book, except that the system has certain special qualities, one of which is that it is conservative, both in the sense of conserving energy since the Hamiltonian function does not depend explicitly on time and also in the sense of conserving state space volume. The conservation of energy follows from $\dot{H} = \dot{v}\partial H/\partial v + \dot{x}\partial H/\partial x + \ldots = \dot{v}\dot{x} - \dot{x}\dot{v} + \ldots = 0$, and the conservation of state space volume follows from $\dot{V}/V = \partial^2 H/\partial x \partial v - \partial^2 H \partial v \partial x + \ldots = 0$. As a consequence, the dynamics occur on a hypersurface of constant energy with a dimension one less than the dimension of the state space as determined by the initial conditions, and there is no attractor. A cluster of initial conditions conserves its volume (Liouville's theorem), and the flow is incompressible and time-reversible. The Lyapunov exponents occur in equal and opposite pairs and thus sum to zero.

6.4.1 *Coupled nonlinear oscillators*

The Hamiltonian functions for each of the coupled oscillators in Table 6.3 are given respectively by

$$CO_0 : H = \frac{v^2}{2} + \frac{u^2}{2} - \cos x - \cos y + \frac{(x-y)^2}{2}$$

$$CO_1 : H = \frac{v^2}{2} + \frac{u^2}{2} - \cos x + |y| + \frac{(x-y)^2}{2}$$

$$CO_2 : H = \frac{v^2}{2} + \frac{u^2}{2} - \cos x + \frac{y^4}{4} + \frac{(x-y)^2}{2}$$

$$CO_3 : H = \frac{v^2}{2} + \frac{u^2}{2} + |x| + |y| + \frac{(x-y)^2}{2}$$ (6.13)

$$CO_4 : H = \frac{v^2}{2} + \frac{u^2}{2} + |x| + \frac{y^4}{4} + \frac{(x-y)^2}{2}$$

$$CO_5 : H = \frac{v^2}{2} + \frac{u^2}{2} + \frac{x^4}{4} + \frac{y^4}{4} + \frac{(x-y)^2}{2}.$$

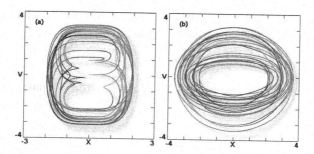

Fig. 6.10 State space plots for two identical velocity coupled oscillators (a) linearly coupled nonlinear oscillators from Eq. (6.15) with $k = 0.5$ and initial conditions $(x_0, v_0, y_0, u_0) = (2, 0, 0, 0)$, $\lambda = (0.1290, 0, 0, -0.1290)$, and (b) nonlinearly coupled linear oscillators from Eq. (6.17) with $k = 1$ and initial conditions $(x_0, v_0, y_0, u_0) = (1, 2, 3, 0)$, $\lambda = (0.3392, 0, 0, -0.3392)$.

Each of these cases consists of two terms $v^2/2 + u^2/2$ that represent the kinetic energy of the oscillators, two terms that represent the potential energy of the oscillators and that depend on x and y alone, and a coupling term $(x - y)^2/2$ that represents the potential energy stored in the hypothetical spring that couples the two oscillators. A useful exercise is to show that Hamilton's equations applied to these Hamiltonians does produce the systems in Table 6.3.

6.4.2 *Velocity coupled oscillators*

Oscillators can also be coupled through their velocities, a simple example of which using two nonlinear (cubic) oscillators is given by the Hamiltonian

$$H = \frac{1}{2}(v^2 + u^2) + \frac{1}{4}(x^4 + y^4) + kvu \qquad (6.14)$$

where k is the coupling constant. Such a coupling would be unusual in a mechanical system but could be easily achieved in an electrical circuit. The corresponding dynamical equations are

$$
\begin{aligned}
\dot{x} &= v + ku \\
\dot{v} &= -x^3 \\
\dot{y} &= u + kv \\
\dot{u} &= -y^3,
\end{aligned}
\qquad (6.15)
$$

which have chaotic solutions for $0 < k < 1$. The state space plot for the case with $k = 0.5$ is shown in Fig. 6.10(a).

It is also possible for the coupling to be nonlinear, in which case chaos can occur even when the oscillators are linear as given by the Hamiltonian

$$H = \frac{1}{2} \left[v^2 + u^2 + x^2 + y^2 + k(vu)^2 \right] \tag{6.16}$$

where k is the coupling constant. The corresponding dynamical equations are

$$
\begin{aligned}
\dot{x} &= v + kvu^2 \\
\dot{v} &= -x \\
\dot{y} &= u + kuv^2 \\
\dot{u} &= -y,
\end{aligned}
\tag{6.17}
$$

which have chaotic solutions for $k = 1$. The state space plot for this case is shown in Fig. 6.10(b). Note that if one simply reinterprets v and u as displacements and x and y as velocities and changes a sign, this system is then force-coupled through a rather odd combination of its component displacements.

6.4.3 *Parametrically coupled oscillators*

Two linear oscillators can also be parametrically coupled, for example using the Hamiltonian

$$H = \frac{1}{2} \left[v^2 + u^2 + x^2 + y^2 + k(xy)^2 \right] \tag{6.18}$$

whose similarity to Eq. (6.16) is evident. The corresponding dynamical system can be written as

$$
\begin{aligned}
\ddot{x} &= -(1 + ky^2)x \\
\ddot{y} &= -(1 + kx^2)y,
\end{aligned}
\tag{6.19}
$$

which has chaotic solutions for $k = 1$ with a state space plot as shown in Fig. 6.11, which is a rotated variant of Fig. 6.10(b). This system is identical to Eq. (6.17) but with the roles of displacement and velocity interchanged, and the different Lyapunov exponents are a consequence of the different initial conditions which determine the energy in the system.

6.4.4 *Simplest Hamiltonian*

A useful method for inventing new four-dimensional conservative chaotic systems is to specify a Hamiltonian $H = H(x, v, y, u)$ and apply Hamilton's equations. There are many combinations of oscillators and couplings that

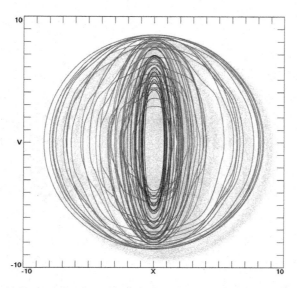

Fig. 6.11 State space plot for two parametrically coupled linear oscillators in Eq. (6.19) with $k = 1$ and initial conditions $(x_0, v_0, y_0, u_0) = (6, 0, 1, 1)$, $\lambda = (0.8565, 0, 0, -0.8565)$.

produce chaos, only a small sample of which were given above. However, it is interesting to ask what is the simplest polynomial Hamiltonian that will produce chaos. It must be at least cubic so that the resulting equations have a quadratic nonlinearity, and it must involve at least four variables since the constancy of the Hamiltonian constrains the dynamics to occur in a space with a dimension one less than the dimension of the state space.

Perhaps the simplest such example is

$$H = v^2 - u^2 - 2y^2 - x^2y, \qquad (6.20)$$

which from Hamilton's equations leads to the dynamical system

$$\begin{aligned}
\dot{x} &= 2v \\
\dot{v} &= 2xy \\
\dot{y} &= -2u \\
\dot{u} &= 4y + x^2.
\end{aligned} \qquad (6.21)$$

Rescaling the variables according to $x \to \sqrt{8}x, y \to 2y, t \to t/\sqrt{8}$, gives a system that can be written in compact form as

$$\begin{aligned}
\ddot{x} &= xy \\
\ddot{y} &= -y - x^2.
\end{aligned} \qquad (6.22)$$

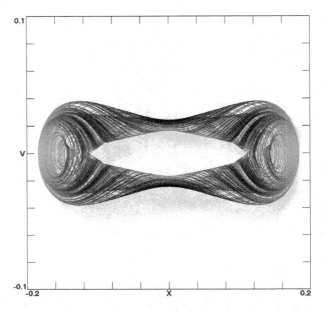

Fig. 6.12 State space plot for Eq. (6.22) derived from a rescaling of the simplest polynomial Hamiltonian in Eq. (6.20) with initial conditions $(x_0, \dot{x}_0, y_0, \dot{y}_0) = (0.04, 0.02, -0.04, 0.11)$, $\lambda = (0.0037, 0, 0, -0.0037)$.

The state space plot for this case is shown in Fig. 6.12. Since Eq. (6.21) has only five terms and two quadratic nonlinearities, it is in some sense as simple as the three-dimensional system in Eq. (4.1), albeit with one additional variable.

6.4.5 *Hénon–Heiles system*

Equation (6.22) is a simplified version of a system developed by Hénon and Heiles (1964) as an approximation to the motion of stars orbiting in a plane about the galactic center with a Hamiltonian

$$H = \frac{1}{2}\left(v^2 + u^2 + x^2 + y^2\right) + x^2 y - \frac{1}{3}y^3 \tag{6.23}$$

and corresponding equations of motion given by

$$\begin{aligned} \ddot{x} &= -x - 2xy \\ \ddot{y} &= -y - x^2 + y^2. \end{aligned} \tag{6.24}$$

Apart from a sign change, this system reduces to Eq. (6.24) if the $-x$ term in the first equation and the y^2 term in the second equation are removed.

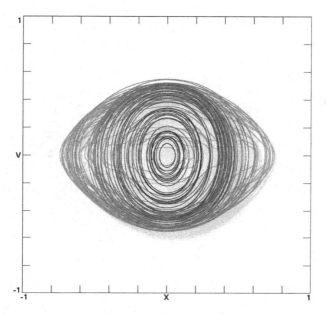

Fig. 6.13 State space plot for the Hénon–Heiles system in Eq. (6.24) with initial conditions $(x_0, v_0, y_0, u_0) = (0.56, 0, 0, 0)$, $\lambda = (0.1045, 0, 0, -0.1045)$.

This *Hénon–Heiles system* is chaotic for $1/12 < H < 1/6$ with a typical state space plot as shown in Fig. 6.13 for $H = 0.1568$. This system provided one of the early observations of what we now call 'chaos' in the numerical solution of a simple system of ODEs, and the authors remarked on their surprise at observing an apparently 'random' distribution of points in the Poincaré section.

6.4.6 *Reduced Hénon–Heiles system*

It turns out that the linear terms are not essential for obtaining chaos in the Hénon–Heiles system, and one can remove them using the Hamiltonian

$$H = -\frac{1}{2}\left(v^2 + u^2 + x^2 y\right) - 0.19 y^3, \tag{6.25}$$

which leads to the simple system

$$\begin{aligned} \ddot{x} &= xy \\ \ddot{y} &= -x^2 + 0.57 y^2 \end{aligned} \tag{6.26}$$

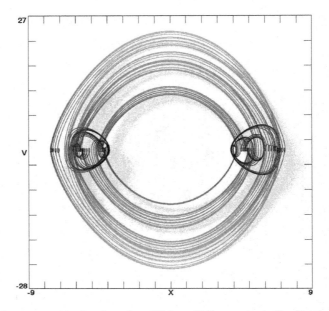

Fig. 6.14   State space plot for the reduced Hénon–Heiles system in Eq. (6.26) with initial
conditions $(x_0, v_0, y_0, u_0) = (3.92, 1.31, 2.04, 1.06)$, $\lambda = (0.0015, 0, 0, -0.0015)$.

whose state space plot is shown in Fig. 6.14. This is a system comparable in
elegance to Eq. (6.22) but with three quadratic nonlinearities and a rather
awkward parameter (0.57).

6.4.7   *N-body gravitational systems*

One of the earliest systems for which Hamiltonian dynamics was applied
is the problem of N bodies interacting gravitationally, such as in the Solar
System. Celestial motion represents one of the few physical examples in
which friction is negligible. The Hamiltonian for such a system is

$$H = \sum_{i=1}^{N} \left(\frac{p_i^2}{2} - m_i \sum_{\substack{j=1 \\ j \neq i}}^{N} \frac{m_j}{R_{ij}} \right) \tag{6.27}$$

where $p_i = m_i \sqrt{v_i^2 + u_i^2 + w_i^2}$ is the momentum of the ith body, $v_i =
\dot{x}_i/m_i, u_i = \dot{y}_i/m_i, w_i = \dot{z}_i/m_i, R_{ij} = \sqrt{(x_i - x_j)^2 + (y_i - y_j)^2 + (z_i - z_j)^2}$
is the distance between body i and body j, and m_i is the mass of body

i in normalized units such that the gravitational constant $G = 1$. The corresponding equations of motion are

$$\ddot{x}_i = \sum_{\substack{j=1 \\ j \neq i}}^{N} m_j \frac{x_j - x_i}{R_{ij}^3}$$

$$\ddot{y}_i = \sum_{\substack{j=1 \\ j \neq i}}^{N} m_j \frac{y_j - y_i}{R_{ij}^3} \qquad (6.28)$$

$$\ddot{z}_i = \sum_{\substack{j=1 \\ j \neq i}}^{N} m_j \frac{z_j - z_i}{R_{ij}^3}.$$

The R^3 in the denominator is not a mistake despite the $1/R^2$ force since an additional power of R is required to normalize the component of the displacement in the numerator in each of the three spatial directions. These equations are much less elegant than most of the other examples in this book, but they are included because of their historical importance in the development of theoretical physics and because there are some rather simple cases for which chaos occurs.

The case with $N = 2$ (for example, a single planet orbiting a single star) was understood by Johannes Kepler who in 1605 announced that the planets in the Solar System moved in elliptical orbits with the Sun at one focus of the ellipse. Isaac Newton subsequently proved from his laws of motion that any two objects gravitationally bound to one another will orbit in ellipses about their center of mass. There are also unbounded hyperbolic orbits, for example where an object approaches the Sun from deep space and is deflected by it, never to return, but we are concerned here only with bounded solutions.

The case with $N = 3$ has been studied by many people including Newton and especially the famous French mathematician Henri Poincaré, who won a prize in 1889 offered by King Oscar II of Sweden for demonstrating that the problem was impossible to solve (Poincaré, 1890). Much subsequent work has been done on the three-body problem, but questions remain. Even less is known about the case with $N > 3$, and there is still debate about the stability of the Solar System (Laskar, 1989; Sussman and Wisdom, 1992), which was the original challenge offered by King Oscar.

6.4.7.1   *Three-body problem*

The simplest case that admits chaos is the so-called *three-body problem* ($N = 3$), which is a system of 18 coupled ODEs (Marchal, 1990; Gutzwiller, 1998; Valtonen and Karttunen, 2006). One simplification is to take the bodies identical with masses equal to 1.0, in which case there are no adjustable parameters except for the initial conditions. Without loss of generality, one can calculate the position and velocities of the bodies in the center-of-mass coordinate system, which amounts to a constraint on the positions $\sum x_i = \sum y_i = \sum z_i = 0$ and velocities $\sum v_i = \sum u_i = \sum w_i = 0$. The remaining system then has twelve independent equations that uniquely specify its dynamics. Some typical state space plots are shown in Fig. 6.15. Only the trajectory of the first body is shown in the xy-plane to keep the plots relatively uncluttered.

6.4.7.2   *Restricted three-body problem*

Another important special case with $N = 3$ is the *restricted three-body problem* (also called *Euler's three-body problem*) in which one of the masses is negligible ($m_1 = 0$) and orbits the other two without significantly affecting their motion, such as a tiny planet in a binary star system. It could also represent the classical limit (ignoring quantum effects) of an electron orbiting two protons in the singly-charged hydrogen molecule, H_2^+ (Pauli, 1922). Without loss of generality, the two stars (or protons) can be taken to orbit in the xy-plane with their center of mass at rest at the origin ($x = y = 0$). If we assume the stars have equal unit masses, $m_2 = m_3 = 1$, then it suffices to calculate the orbit of only one of the stars since the other will always have equal and opposite position and velocity. In addition, if we assume that the planet also remains in the xy-plane, the equations of motion simplify to

$$
\begin{aligned}
\ddot{x}_1 &= (x_2 - x_1)/R_{12}^3 - (x_2 + x_1)/R_{13}^3 \\
\ddot{y}_1 &= (y_2 - y_1)/R_{12}^3 - (y_2 + y_1)/R_{13}^3 \\
\ddot{x}_2 &= -2x_2/R_{23}^3 \\
\ddot{y}_2 &= -2y_2/R_{23}^3.
\end{aligned}
\tag{6.29}
$$

This eight-dimensional system is probably the simplest three-body case that admits chaotic solutions.

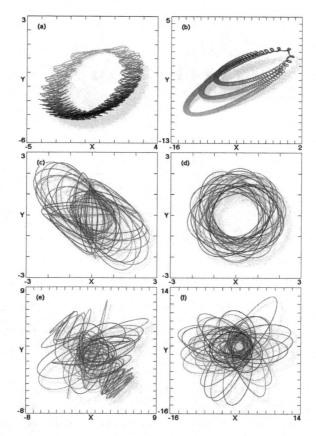

Fig. 6.15   State space plots for one of three equal masses interacting gravitationally by Eq. (6.28) with initial conditions $(x_{10}, y_{10}, z_{10}, v_{10}, u_{10}, w_{10}, x_{20}, y_{20}, z_{20}, v_{20}, u_{20}, w_{20})$ (a) $(-2, -1, -1.6, 0, 0.2, 0, -1.3, 0, -0.1, 0.2, -0.5, 0.4), \lambda = 0.0114$, (b) $(0, 0, 0, 0, -0.9, -1, -1, 0, 0, 0, 0, 0), \lambda = 0.0105$, (c) $(0, 0, -1, 0, -1, 0, 0, 0, 0, 1, 1, 0), \lambda = 0.1681$, (d) $(-1, 0, -0.5, 0, -1, 0, 0, 0, 1, 0, 0, 0), \lambda = 0.3344$, (e) $(1, 0, 1, 0, 0, 1, -1, 1, 0, 0, 0, 0), \lambda = 0.0529$, (f) $(0, -1, 1, -1, 0, 0, 0, 0, -2, 0, 0, 0), \lambda = 0.0255$.

Since there are no adjustable parameters, the dynamics are controlled entirely by the set of eight initial conditions (four initial positions and four initial velocities). Some sample chaotic orbits are shown in Fig. 6.16. In this case, all three bodies are shown since the stars orbit in simple symmetric ellipses centered on the origin, unperturbed by the planet. In cases (a) and (b), the planet orbits both stars with an orbit that approximates a precessing ellipse, and in cases (c) and (d) the planet orbits one of the stars with its orbit perturbed by the other star.

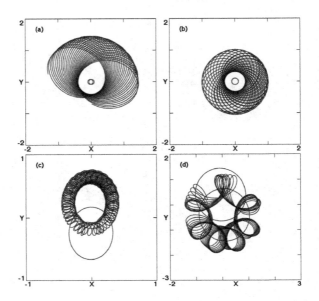

Fig. 6.16   State pace plots for a tiny planet orbiting a pair of identical stars in a plane according to Eq. (6.29) with initial conditions $(x_{10}, y_{10}, v_{10}, u_{10}, x_{20}, y_{20}, v_{20}, u_{20})$ (a) $(-0.3, -1.4, -0.8, 0, 0.1, 0, 0, 1.4), \lambda = 0.0067$, (b) $(0, 1, 1, 0, 0.1, 0, 0, 1.6), \lambda = 0.0139$, (c) $(-0.4, 0.3, -0.9, 1.4, 0.3, -0.4, 0.2, 0.6), \lambda = 0.0208$, (d) $(1, 0, 1, 0.5, 1, -0.7, 0.1, 0.4)$, $\lambda = 0.0047$.

This system is essentially a master-slave oscillator. The last two equations in Eq. (6.29) are decoupled from the first two. A planet orbiting a single star is a nonlinear oscillator without forcing and thus with periodic solutions, whereas a planet orbiting a pair of stars is a nonlinear oscillator forced by the periodic orbital motion of the stars around one another. Thus it is not surprising that it has chaotic solutions. It is interesting to contemplate whether such a planet would be a good candidate for extraterrestrial life and what it would be like to live there.

6.4.8   *N-body Coulomb systems*

Closely related to the N-body gravitational system is the *N-body Coulomb system* in which the forces between the bodies obey an inverse square law but that can be either attractive, if their charges have opposite signs, or repulsive if their charges have the same signs (either both positive or both negative). In the limit of very large N with a nearly equal number of positive and negative charges, such a system could represent a quasi-neutral plasma (Chen, 1984). Although much of the interest in this problem has been in the quantum mechanical limit describing multi-electron atoms such as helium and collisions of charged particles with atoms (Lin, 1995; Tanner *et al.*, 2000), we will consider only the classical, nonrelativistic case.

6.4.8.1 *Three spatial dimensions*

The equations of motion of such a Hamiltonian system are

$$\ddot{x}_i = \sum_{\substack{j=1 \\ j \neq i}}^{N} \frac{q_i q_j}{m_i} \frac{x_i - x_j}{R_{ij}^{\gamma}}$$

$$\ddot{y}_i = \sum_{\substack{j=1 \\ j \neq i}}^{N} \frac{q_i q_j}{m_i} \frac{y_i - y_j}{R_{ij}^{\gamma}} \qquad (6.30)$$

$$\ddot{z}_i = \sum_{\substack{j=1 \\ j \neq i}}^{N} \frac{q_i q_j}{m_i} \frac{z_i - z_j}{R_{ij}^{\gamma}}$$

where q_i is the charge of the ith particle and m_i is its mass. For the usual $1/R^2$ Coulomb force between two point charges in three spatial dimensions, γ is equal to 3. Earnshaw's theorem (Earnshaw, 1842) states that a collection of point charges cannot be maintained in a stable stationary equilibrium solely by their electrostatic forces, but it does not preclude dynamical solutions.

The case with $N = 2$ is relatively uninteresting. If the charges have opposite signs, the force is attractive, and the behavior is identical to the two-body gravitational problem with the bodies either moving in elliptical orbits about their center of mass or moving to infinity along hyperbolic orbits. If the charges have the same sign, the force is repulsive, and the bodies move off to infinity in opposite directions relative to their center of mass, which itself may be moving. Similarly, the case with $N > 2$ is unbounded if the charges all have the same sign.

The three-body Coulomb problem with unequal charges is one of the oldest unsolved problems in theoretical physics. We will not solve it here, but rather will look for chaotic solutions in the case where one of the charges is opposite in sign to the other two. As with the gravitational case, with-

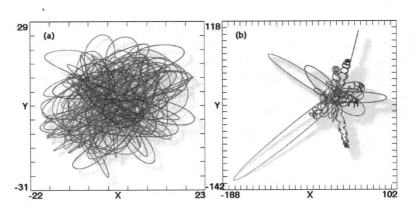

Fig. 6.17 Orbit of ody 1 for the three-body, three-dimensional, Coulomb system in Eq. (6.30) (a) with $(m_1, m_2, m_3, q_1, q_2, q_3, \gamma) = (1, 1, 1, -2, 1, 1, 1)$ and initial conditions $(x_{10}, y_{10}, z_{10}, \dot{x}_{10}, \dot{y}_{10}, \dot{z}_{10}, x_{20}, y_{20}, z_{20}, \dot{x}_{20}, \dot{y}_{20}, \dot{z}_{20}) = (0, -7, -5.3, 2.4, -8, 0, 6, 7.2, 22, -0.3, 1.1, 1)$, $\lambda = 0.0294$ and (b) with $(m_1, m_2, m_3, q_1, q_2, q_3, \gamma) = (1, 1, 1, -1, 4, -1, 2)$ and initial conditions $(x_{10}, y_{10}, z_{10}, \dot{x}_{10}, \dot{y}_{10}, \dot{z}_{10}, x_{20}, y_{20}, z_{20}, \dot{x}_{20}, \dot{y}_{20}, \dot{z}_{20}) = (9.9, -8, -0.3, 0.1, 1, 0, 1, 9, 0.9, 1.1, 1.4, 0)$, $\lambda = 0.0248$.

out loss of generality, we can take the center of mass to be at rest at the origin, which eliminates six of the eighteen equations since $m_3 x_3 = -m_1 x_1 - m_2 x_2, m_3 \dot{x}_3 = -m_1 \dot{x}_1 - m_2 \dot{x}_2$, and similarly for the y and z components.

The usual case with $\gamma = 3$ (corresponding to a $1/R^2$ force) appears not to have bounded solutions for any (nonzero) choice of masses and charges. However, other values of γ do have bounded, and in fact chaotic solutions, one example of which with $\gamma = 1$ (corresponding to a force that is independent of separation) is shown in Fig. 6.17(a). The figure shows the motion of body 1 in the xy-plane. The motion of the other two bodies is similar. Apparently such solutions exist for all masses equal $(m_1 = m_2 = m_3)$, but not for equal magnitudes of the charges as suggested by an extensive unsuccessful search for such cases.

There are also chaotic solutions with $\gamma = 2$ (corresponding to a $1/R$ force) with unequal charges, one example of which is shown in Fig. 6.17(b).

6.4.8.2 *Two spatial dimensions*

An even simpler case constrains the three bodies to move only in the xy-plane. The resulting eight-dimensional system has chaotic solutions, one

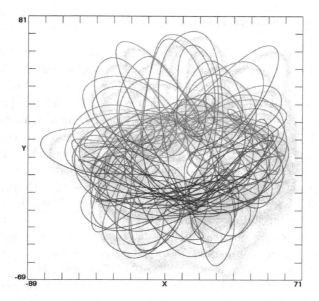

Fig. 6.18 Orbit of ody 1 for the three-body, two-dimensional, Coulomb system in Eq. (6.31) with initial conditions $(x_{10}, y_{10}, \dot{x}_{10}, \dot{y}_{10}, x_{20}, y_{20}, \dot{x}_{20}, \dot{y}_{20}) = (0, 52, -5.7, 0, 10, -52.8, 0, -0.8)$, $\lambda = 0.0276$.

simple example of which with all three masses equal is given by

$$\ddot{x}_1 = \frac{x_1 - x_2}{R_{12}} - 2\frac{x_1 - x_3}{R_{13}}$$

$$\ddot{y}_1 = \frac{y_1 - y_2}{R_{12}} - 2\frac{y_1 - y_3}{R_{13}}$$

$$\ddot{x}_2 = \frac{x_2 - x_1}{R_{12}} - 2\frac{x_2 - x_3}{R_{23}} \qquad (6.31)$$

$$\ddot{y}_2 = \frac{y_2 - y_1}{R_{12}} - 2\frac{y_2 - y_3}{R_{23}}$$

with $x_3 = -x_1 - x_2$ and $y_3 = -y_1 - y_2$ as required for conservation of momentum. In terms of Eq. (6-30), the respective charges are $(q_1, q_2, q_3) = (-1, -1, 2)$ and the force law is $\gamma = 1$, corresponding to a force that is independent of separation. The orbit of body 1 in the xy-plane is shown in Fig. 6.18. In this figure, the third dimension, as illustrated by the gray level and shadow, is proportional to $\dot{x}_1$.

6.5 Anti-Newtonian Systems

Forces that obey Newton's laws are obviously important, but they represent a small subclass of possible dynamics. Imagine instead a world in which Newton's first and second laws hold but where Newton's third law takes the form that the forces between any two objects are equal in both magnitude *and* direction. Such systems might be called 'anti-Newtonian' (Sprott, 2009), although that term has also been used to describe a general relativistic universe with a purely gravito-magnetic field (Maartens *et al.*, 1998). Equivalently, one could consider one of the bodies to have a negative inertial mass (Bondi, 1957), in which case Newton's third law would hold, but the body with a negative mass would accelerate in a direction opposite to the applied force and would have a negative kinetic energy, which is conceptually unappealing. Violations of Newton's third law are not uncommon, for example in the forces between moving charges when retardation effects are considered (Keller, 1942; Gerjuoy, 1949). However, an anti-Newtonian force pair is admittedly unusual, but it might approximate some biological process such as a spatial predator–prey problem in which the fox is attracted to the rabbit, but the rabbit is repelled by the fox.

6.5.1 *Two-body problem*

The simplest example would consist of only two such objects interacting according to these peculiar rules. Unlike the Newtonian and Coulomb cases, two bodies interacting in this way do not conserve energy or momentum since one body pulls the other while the latter pushes the former, adding energy to both. Consequently, the center of mass does not move with a constant velocity. In the biological example, the energy comes from metabolism, and the momentum comes from friction with the ground along which they are running. As a result, peculiar things can happen such as both bodies rebounding with infinite velocity after a collision. More interestingly, the absence of conservation laws means that chaos can occur in systems with fewer state space dimensions than would otherwise be required, and such systems can have nontrivial attractors in the presence of dissipation because there is a source of energy inherent in the interaction. Such dissipative systems are not Hamiltonian.

For two bodies free to move in only one spatial dimension (x), solutions are usually unbounded in the absence of dissipation unless at least one of the bodies has infinite mass, in which case bounded oscillations can occur if the objects are penetrable. The infinitely massive object absorbs the momentum and energy generated in the interaction without moving. In the presence of dissipation (friction), stable limit cycles can occur even for bodies of equal mass, for example in the system $\ddot{x}_1 = \text{sgn}(x_1-x_2)-2\dot{x}_1, \ddot{x}_2 = \text{sgn}(x_1 - x_2) - \dot{x}_2$ where the rabbit (body 1) and fox (body 2) oscillate back and forth with the fox continually overshooting the rabbit, causing them both to reverse direction. The positive work done by the interaction force just balances the energy lost through friction when averaged over a cycle of the oscillation.

In two dimensions without dissipation, bounded solutions are possible only if the average net work done on the bodies is zero. The simplest way to arrange this is to have the bodies orbit synchronously in concentric circles so that the force is radial and hence perpendicular to the velocity. For example, if the rabbit has twice the mass of the fox (perhaps it is a lamb rather than a rabbit) and orbits at half the radius, then the forces on them are equal if they orbit with the same angular velocity. This result is independent of how the force depends on their separation since the separation is constant.

However, there is another solution with more complicated orbits in which the positive work done on the system when the fox is in pursuit of the rabbit is just balanced by the negative work done after the fox overshoots the rabbit. One such example has the rabbit twice as massive as the fox and the force inversely proportional to their separation, which gives a quasi-periodic trajectory with the rabbit and fox in synchronous, precessing orbits. A relatively weak force law such as $1/R$ is not unreasonable for such a biological example, where the attraction/repulsion is not strongly dependent on the separation as long as the two individuals are within line of sight. An animated version of this case along with other anti-Newtonian examples is available at `http://sprott.physics.wisc.edu/antinewt.htm`.

The simplest anti-Newtonian case for which chaotic solutions are obtained involves two bodies moving in a plane with dissipation, one simple example of which is given by

$$\ddot{x}_1 = 2(x_1 - x_2)/R^2 - 4\dot{x}_1$$

$$\ddot{y}_1 = 2(y_1 - y_2)/R^2 - 4\dot{y}_1$$

$$\ddot{x}_2 = (x_1 - x_2)/R^2 - \dot{x}_2 \tag{6.32}$$

$$\ddot{y}_2 = (y_1 - y_2)/R^2 - \dot{y}_2$$

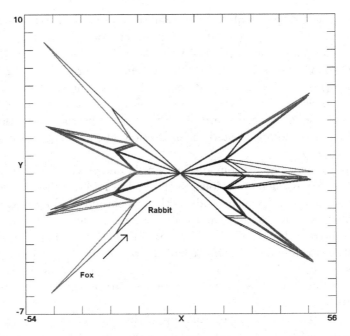

Fig. 6.19 Attractor for the two-body anti-Newtonian system in Eq. (6.32) with $(x_{10}, y_{10}, \dot{x}_{10}, \dot{y}_{10}, x_{20}, y_{20}, \dot{x}_{20}, \dot{y}_{20}) = (0.3, 24, 0, 0, -0.2, 47, 0, 0)$, $\lambda = 0.0453$.

where $R = \sqrt{(x_1 - x_2)^2 + (y_1 - y_2)^2}$ is the Euclidean distance between the bodies. Note that body 1 (the rabbit) has half the mass and twice the friction as body 2 (the fox) and that the force is proportional to $1/R$ in contrast to the $1/R^2$ for the gravitational case previously described. The orbits of the two bodies in the xy-plane are shown in Fig. 6.19. Because of the singularity in the force, calculation of the orbit is computationally difficult, and the quoted Lyapunov exponent is only an approximation since its value is dominated by the dynamics during the relatively close encounters of the bodies.

This unusual attractor deserves some comment. To a first approximation, the rabbit and fox are oscillating back and forth along a line as in the case with one spatial dimension. The fox overtakes the rabbit on the outbound leg but misses slightly, passing either to the right or left of the rabbit by a nearly constant distance of $R \approx 0.01$. Then they both slow nearly to a halt and resume the chase in the opposite direction, but rotated by a bit less than 6 degrees in the plane because of their nearly equal deflection during the close encounter. The trajectory would exhibit a regular precession except for the fact that the sequence of right and left passings is apparently chaotic and has properties indistinguishable from a sequence of coin tosses, causing the orbit to walk randomly in angle. That makes the system ripe for analysis by symbolic dynamics (Lind and Marcus, 1995; Kitchens, 1998). A typical sequence of passings is given by

LLLRLLRRLLLRLRRRRRLLRLRRLLRLLRLLRLRRRLLRRRRLLLLL
LRLLRLLRLLRLRRLRRRLLRLLLLLLLLRLRRRLRRLLLLRRRLRLRL
RLLLLLLRRRRRRRLLLLLLLLRRRLRRLLLLRRRLLLLRRLLRLLRL
RLLLLRLLLRLRLRLLRRLLLRLRRLRLLLRRLLLRRRLLLLLRRRRR
LLRRLLLRLLRLLLRRRRLR,

which has no discernable pattern.

6.5.2   *Three-body problem*

Just as with the gravitational and Coulomb cases, anti-Newtonian systems can also involve three bodies. Two of the bodies must be either attracting (rabbits) or repelling (foxes), and we will assume that the bodies of the same type do not interact. In two dimensions without dissipation, bounded solutions are possible if the three bodies orbit synchronously in concentric circles so that the forces are radial and hence perpendicular to the velocity. For example, two foxes can circle the one rabbit either both in phase or one in phase and the other 180 degrees out of phase with the rabbit if the radii are chosen appropriately. Similarly, one fox can circle two rabbits with all three in phase provided the fox has sufficient mass and orbits outside both rabbits. This result is independent of how the force depends on their separation since the separations are constant.

With dissipation, the dynamics are more varied. Because of the asymmetry between the bodies, there are two possibilities — one fox chasing two rabbits, or two foxes chasing one rabbit. It turns out that in the former case, at least one of the rabbits will usually escape to infinity while the fox is preoccupied with chasing the other rabbit, and the problem reduces to the two-body case previously considered, provided the forces decrease sufficiently rapidly with separation.

For the case of two foxes chasing one rabbit, chaotic solutions are possible, even when the foxes have identical masses and the same friction, one

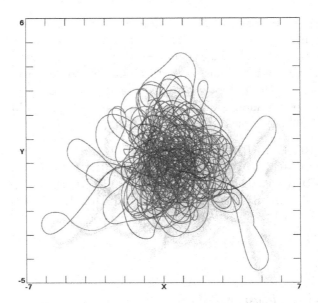

Fig. 6.20   Attractor showing the trajectory of a rabbit being chased by two foxes in the three-body anti-Newtonian system in Eq. (6.33) with $(x_{10}, y_{10}, \dot{x}_{10}, \dot{y}_{10}, x_{20}, y_{20}, \dot{x}_{20}, \dot{y}_{20},$ $x_{30}, y_{30}, \dot{x}_{30}, \dot{y}_{30}) = (0.5, 1, 0, 0, 1, 0.1, -0.4, 0.4, 0, 2, 0.4, 0)$, $\lambda = 0.1346$.

example of which is given by

$$\ddot{x}_1 = (x_1 - x_2)/R_{12}^2 + (x_1 - x_3)/R_{13}^2 - 3\dot{x}_1$$

$$\ddot{y}_1 = (y_1 - y_2)/R_{12}^2 + (y_1 - y_3)/R_{13}^2 - 3\dot{y}_1$$

$$\ddot{x}_2 = 0.5\left[(x_1 - x_2)/R_{12}^2 - \dot{x}_2\right]$$

$$\ddot{y}_2 = 0.5\left[(y_1 - y_2)/R_{12}^2 - \dot{y}_2\right] \tag{6.33}$$

$$\ddot{x}_3 = 0.5\left[(x_1 - x_3)/R_{13}^2 - \dot{x}_3\right]$$

$$\ddot{y}_3 = 0.5\left[(y_1 - y_3)/R_{13}^2 - \dot{y}_3\right]$$

where the foxes are assumed not to interact with one another. This system has a twelve-dimensional state space, and a plot of the rabbit's trajectory in the xy-plane as shown in Fig. 6.20 illustrates the complexity of the solution. The trajectories of the foxes are similar. An animated version of this case is available at http://sprott.physics.wisc.edu/chaos/zhdankin/2foxes.htm.

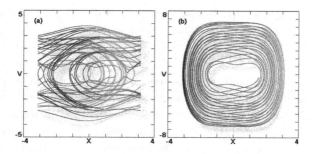

Fig. 6.21 State pace plots for some periodically forced snap oscillators (a) from Eq. (6.34) with $(x_0, \dot{x}_0, \ddot{x}_0, \dddot{x}_0) = (1, 0, -\sin 1, 1), \lambda = (0.1484, 0, 0, -0.1484)$, (b) from Eq. (6.35) with $(x_0, \dot{x}_0, \ddot{x}_0, \dddot{x}_0) = (0, 0, 0, 2), \lambda = (0.0910, 0, 0, -0.0910)$.

6.6 Hyperjerk Systems

Many three-dimensional systems of ODEs can be written in jerk form as indicated by Eq. (3.5), suggesting that a similar transformation might be possible in higher dimensions. Whereas the first three derivatives ($\dot{x}, \ddot{x}$, and $\dddot{x}$) have well accepted names (*velocity*, *acceleration*, and *jerk*, respectively), there is no universally accepted name for the higher derivatives. The fourth derivative ($\ddddot{x}$) has been called a 'jounce,' a 'sprite,' and a 'surge,' but perhaps the best name is a 'snap' with successive derivatives 'crackle' and 'pop' (Sprott, 1997b). Generically, we will refer to systems with time derivatives higher than the third as *hyperjerk systems* (Chlouverakis and Sprott, 2006), and we will refer to ones that can be written as $\ddddot{x} = f(\dddot{x}, \ddot{x}, \dot{x}, x)$ as *snap systems*, and similarly for *crackle* and *pop*.

6.6.1 *Forced oscillators*

All the periodically forced oscillators and many of the coupled oscillators previously discussed can be cast into snap form (Linz, 2008). For example, the frictionless forced pendulum (Model FO_0 in Table 6.1) in Eq. (6.2) can be written as

$$\ddddot{x} + (1 + \cos x)\ddot{x} + (1 - \dot{x}^2)\sin x = 0. \tag{6.34}$$

Figure 6.21(a) shows that its state space plot for equivalent initial conditions is the same as Fig. 6.1. Furthermore, the Lyapunov exponents are the same to within numerical precision, lending further credence to the equivalence of the two representations.

Another example is the periodically forced undamped oscillator with a cubic restoring force (Model FO_7 in Table 6.1), which can be written as

$$\dddot{x} + (4 + 3x^2)\ddot{x} + (6\dot{x}^2 + 4x^2)x = 0. \tag{6.35}$$

Figure 6.21(b) shows that its state space plot and Lyapunov exponents for equivalent initial conditions are nearly the same as model FO_7 of Fig. 6.2.

One could continue to show examples of this sort, including ones with damping (Sprott, 1997b), but it should be evident that the hyperjerk representation is not necessarily a simplification other than reducing the system to one involving a single scalar variable. In fact, the examples shown above are two of the simplest such cases. More interesting is to look for the simplest hyperjerk systems that have chaotic solutions.

6.6.2   *Chlouverakis systems*

Chaotic hyperjerk systems have been studied by Chlouverakis and Sprott (2006), who considered systems that involve fourth and fifth derivatives, here called 'snap systems' and 'crackle systems,' respectively.

6.6.2.1   *Snap systems*

One example of such a snap system has the form

$$\ddddot{x} + a\dddot{x} + b\ddot{x} + c\dot{x} = f(x, \dot{x}, \ldots) \tag{6.36}$$

where f is a nonlinear function of x and perhaps its higher derivatives, simple examples of which are polynomials. Chlouverakis and Sprott (2006) reported a number of such cases, one of the simplest of which is a dissipative case given by

$$\ddddot{x} + \dddot{x} + 5.2\ddot{x} + 2.7\dot{x} = 4.5(x^2 - 1) \tag{6.37}$$

whose attractor is shown as model CS_1 in Fig. 6.22.

It turns out that there are even simpler chaotic snaps with a single quadratic nonlinearity such as the conservative system

$$\ddddot{x} + 6\ddot{x} = 1 - x^2 \tag{6.38}$$

whose state space plot is shown as model CS_2 in Fig. 6.22.

Chlouverakis and Sprott (2006) also reported a dissipative case in which the nonlinearity occurs in the $\dot{x}$ term according to

$$\ddddot{x} + \dddot{x} + 4\ddot{x} + \dot{x}^2 + x = 0 \tag{6.39}$$

with an attractor as shown as model CS_3 in Fig. 6.22. This system also

Table 6.4   Chaotic Chlouverakis snap systems.

Model	Equations	$x_0, \dot{x}_0, \ddot{x}_0, \dddot{x}_0$	Lyap Exponents
CS_1	$\ddddot{x} + \dddot{x} + 5.2\ddot{x} + 2.7\dot{x}$ $= 4.5(x^2 - 1)$	$-0.9, 0, 0, 0$	$0.1865, 0,$ $-0.4830, -0.7035$
CS_2	$\ddddot{x} + 6\ddot{x} = 1 - x^2$	$-0.44, 0.13, 0.42, -1.29$	$0.0039, 0,$ $0, -0.0039$
CS_3	$\ddddot{x} + \dddot{x} + 4\ddot{x} + \dot{x}^2$ $+x = 0$	$0.18, 0.48, 0.26, 0.85$	$0.0259, 0,$ $-0.4829, -0.5430$
CS_4	$\ddddot{x} + 8\ddot{x} - \dot{x}^2 + x = 0$	$-1.94, 1.43, -0.98, -1.48$	$0.0055, 0,$ $0, -0.0055$
CS_5	$\ddddot{x} + \dddot{x} + 4\ddot{x} + \dot{x}x = 0$	$2.4, 6, 0, 4$	$0.1633, 0,$ $0, -1.1633$
CS_6	$\ddddot{x} = -2.02\dddot{x} + 2\ddot{x}\dot{x} - \dot{x}$	$4, 2, 0, 0$	$0.1594, 0,$ $-0.0435, -2.1359$

has a conservative counterpart given by

$$\ddddot{x} + 8\ddot{x} - \dot{x}^2 + x = 0 \tag{6.40}$$

whose state space plot is shown as model CS_4 in Fig. 6.22.

A dissipative snap system that is even simpler than Eq. (6.39) but of similar form can be constructed by taking the derivative of each of the terms in model MO_3 from Table 3.3 and simplifying the parameters to obtain an equation of the form

$$\ddddot{x} + \dddot{x} + 4\ddot{x} + \dot{x}x = 0 \tag{6.41}$$

whose attractor is shown as model CS_5 in Fig. 6.22. This system has the curious feature of having two zero Lyapunov exponents, which is unusual for a dissipative system except when the system is at a bifurcation point.

A similar method can be used to convert any of the jerk systems in Chapters 3 and 4 to hyperjerk form, but the resulting systems are dynamically equivalent to the lower-dimensional systems from which they were derived. For example, the simplest quadratic jerk system in Eq. (3.8) (model JD_0 in Table 3.2) can be differentiated to obtain

$$\ddddot{x} = -a\dddot{x} + 2\ddot{x}\dot{x} - \dot{x} \tag{6.42}$$

whose attractor for $a = 2.02$ as shown as model CS_6 in Fig. 6.22 resembles the one in Fig. 3.11 but with rather different Lyapunov exponents. However, since Eq. (6.42) does not contain x, a simple substitution of $y = 2\dot{x}$ reduces it to the three-dimensional jerk system in Eq. (3.9).

The chaotic snap systems studied by Chlouverakis and their generalizations and simplifications are summarized in Table 6.4.

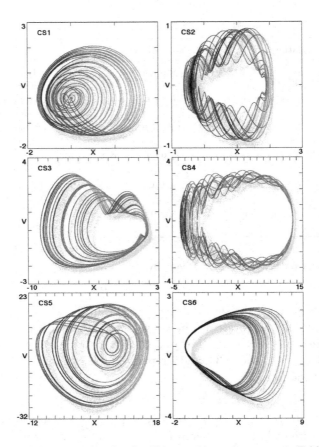

Fig. 6.22   State pace plots for the Chlouverakis snap systems in Table 6.4.

6.6.2.2   *Crackle systems*

A logical extension of these snap systems is to consider systems with a fifth or higher time derivative, and Chlouverakis and Sprott (2006) show some such examples, including ones that exhibit hyperchaos (more shortly). However, it is difficult to find bounded chaotic solutions unless the coefficients are chosen carefully since each successive derivative requires that a corresponding succession of integrals of the hyperjerk function be exactly zero when integrated over an arbitrarily large time. In particular, such systems tend to be unbounded if too many of the intermediate derivatives are omitted. Therefore, such systems tend to be more complicated and delicate, and are thus less elegant than their lower-dimensional counterparts.

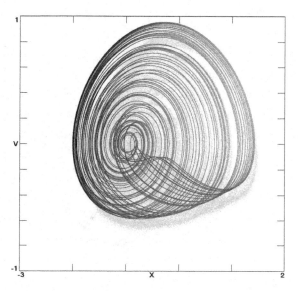

Fig. 6.23   Attractor for the 5-D crackle system in Eq. (6.43) with $(x_0, \dot{x}_0, \ddot{x}_0, \dddot{x}_0, \ddddot{x}_0) =$ $(-0.8, 0, 0, 0, 0)$, $\lambda = (0.0498, 0, -0.2527, -0.3985, -0.3985)$.

These comments notwithstanding, it is possible to find chaotic hyper-jerks with fifth and higher derivatives, especially if the nonlinearity is of the form $x^2 - 1$, one elegant example of which is given by

$$x^{(5)} + \ddddot{x} + 4\dddot{x} + \ddot{x} + \dot{x} = 0.2(x^2 - 1) \tag{6.43}$$

where $x^{(5)} = d^5x/dt^5$ (more than four overdots are hard to read).

The attractor for this case is shown in Fig. 6.23. Because it is a hyperjerk with a fifth derivative, it is called a 'crackle system.'

6.6.2.3   *Pop systems*

There is no reason to stop with the fifth derivative, although the systems become successively more delicate as the order of the derivative increases. An example of a chaotic hyperjerk system with a sixth derivative is given by

$$x^{(6)} + x^{(5)} + 2.6\dddot{x} + 2\ddot{x} + \ddot{x} + 0.5\dot{x} = 0.1(x^2 - 1) \tag{6.44}$$

where $x^{(6)} = d^6x/dt^6$.

The attractor for this case is shown in Fig. 6.24. Because it is a hy-perjerk with a sixth derivative, it is called a 'pop system.' It will be

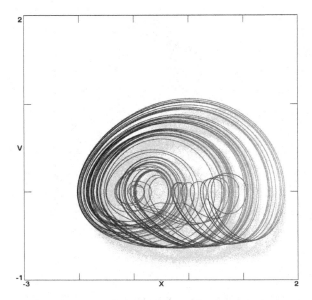

Fig. 6.24   Attractor for the 6-D pop system in Eq. (6.44) with $(x_0, \dot{x}_0, \ddot{x}_0, \dddot{x}_0, \ddddot{x}_0,$
$x_0^{(5)}) = (0.6, 0, 0, 0, 0, 0)$, $\lambda = (0.0528, 0, -0.0784, -0.0813, -0.2954, -0.5978)$.

left as a challenge for the reader to find similar examples with derivatives
higher than the sixth and perhaps to discover rules for their existence and
properties.

6.7   Hyperchaotic Systems

A *hyperchaotic system* is one in which two or more Lyapunov exponents are
positive. Since one exponent must be zero and the sum of the exponents
cannot be positive (else the trajectory would necessarily be unbounded),
hyperchaos can only occur in autonomous systems with four or more state
space variables. Conservative systems with exactly four variables cannot be
hyperchaotic since two of the exponents must be zero and the remaining two
must be equal in magnitude but opposite in sign. Moreover, the strange
attractor that accompanies a dissipative hyperchaotic system must have
a dimension greater than 3.0 since the sum of the first three Lyapunov
exponents must be positive. Thus some of the systems previously discussed
in this chapter are potentially hyperchaotic for an appropriate choice of
their parameters.

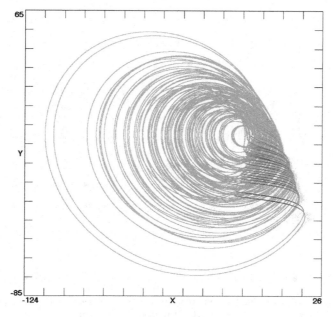

Fig. 6.25 Attractor for the 4-D hyperchaotic Rössler system in Eq. (6.45) with $(a, b, c, d) = (0.25, 3, 0.05, 0.5)$, $(x_0, y_0, z_0, w_0) = (-6, 0, 0.5, 14)$, $\lambda = (0.1120, 0.0211, 0, -24.9312)$.

6.7.1 *Rössler hyperchaos*

One of the first hyperchaotic attractors was discovered by Rössler (1979b) and is given by

$$\begin{aligned}
\dot{x} &= -y - z \\
\dot{y} &= x + ay + w \\
\dot{z} &= b + xz \\
\dot{w} &= cw - dz
\end{aligned} \qquad (6.45)$$

with the parameters $(a, b, c, d) = (0.25, 3, 0.05, 0.5)$, which gives the Lyapunov exponents $\lambda = (0.1120, 0.0211, 0, -24.9312)$ (Wolf *et al.*, 1985). This system is a generalization of the three-dimensional Rössler system in Eq. (3.3) but with an added variable (w) and some added terms. It still serves as the prototypical example of hyperchaos. Its attractor is shown in Fig. 6.25.

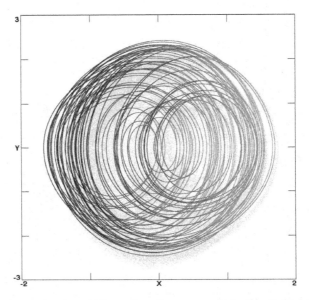

Fig. 6.26   Attractor for the 4-D hyperchaotic snap system in Eq. (6.46) with $A = 3.6$, $(x_0, \dot{x}_0, \ddot{x}_0, \dddot{x}_0) = (0, 0, -2, 0)$, $\lambda = (0.1310, 0.0358, 0, -1.2550)$.

6.7.2   *Snap hyperchaos*

Chlouverakis and Sprott (2006) proposed what may be the algebraically simplest hyperchaotic snap system given by

$$\ddddot{x} + x^4 \dddot{x} + A\ddot{x} + \dot{x} + x = 0 \qquad (6.46)$$

with the single parameter $A = 3.6$ and Lyapunov exponents $\lambda = (0.1310, 0.0358, 0, -1.2550)$. Its attractor is shown in Fig. 6.26.

6.7.3   *Coupled chaotic systems*

A trivial way to achieve hyperchaos in a six-dimensional system is to combine two decoupled, three-dimensional, chaotic, autonomous systems that are not necessarily identical. The spectrum of Lyapunov exponents is then just the combination of their individual Lyapunov exponents. Not surprisingly, a weak coupling between the systems only slightly alters the Lyapunov exponents and thus preserves the hyperchaos, which in many cases persists for relatively large values of the coupling (Cafagna and Grassi, 2003). In fact, it is often the case that one of the zero Lyapunov exponents

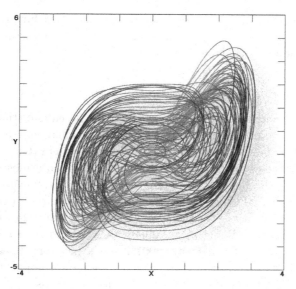

Fig. 6.27 Attractor for the 6-D hyperchaotic system in Eq. (6.47) consisting of two coupled diffusionless Lorenz systems with $k = 0.6$, $R = 1$, and $(x_0, y_0, z_0, u_0, v_0, w_0) = (0, 0.8, 0.4, 0, 1, 0)$, $\lambda = (0.1153, 0.0524, 0, -0.0569, -0.3150, -1.7958)$.

becomes positive when the coupling is weak, leading to a system with three positive Lyapunov exponents (Liu *et al.*, 2003).

A simple example of such a system is two coupled diffusionless Lorenz systems given by

$$
\begin{aligned}
\dot{x} &= y - x + ku \\
\dot{y} &= -xz \\
\dot{z} &= xy - R \\
\dot{u} &= v - u + kx \\
\dot{v} &= -uw \\
\dot{w} &= uv - R.
\end{aligned}
\tag{6.47}
$$

When $k = 0$, this system reduces to two decoupled chaotic systems as in Eq. (3.2), which is chaotic for a range of values near $R = 1$. The system remains hyperchaotic for most values of k up to about $k = 0.6$ for $R = 1$ with an attractor as shown in Fig. 6.27.

Depending on how they are coupled, it may happen that two identical three-dimensional chaotic systems will simply synchronize and behave like a single three-dimensional system, but in the example above and in many similar cases, the synchronized state is unstable, and the dynamics are truly six-dimensional. When the coupling is increased further ($k > 0.6$), the system above becomes periodic with a limit-cycle attractor, albeit with a long-duration chaotic transient.

6.7.4 *Other hyperchaotic systems*

Many other hyperchaotic systems have been developed and studied, usually as extensions of known autonomous three-dimensional chaotic systems including (but by no means limited to) the Lorenz system (Gao *et al.*, 2007), the Lorenz–Haken system (Ning and Haken, 1990), an atmospheric model (Lorenz, 1991), the generalized Lorenz system (Li *et al.*, 2005b), Chua's circuit (Kapitaniak and Chua, 1994), a modified Chua's circuit (Thamilmaran *et al.*, 2004), Chen's system (Li *et al.*, 2005a), Lü's system (Jia, 2007), a modified Lü system (Wang *et al.*, 2006), Qi's system (Qi *et al.*, 2008), a modified Qi system (Chen *et al.*, 2007), and a piecewise-linear snap system (Ahmad, 2006). Most of these systems are more complicated than the hyperchaotic examples above, and none is as elegant as the above hyperchaotic snap system. It appears that hyperchaos is relatively easy to obtain in four-dimensional dynamical systems and even more so as the dimension increases, and thus we will not pursue the matter further here.

6.8 Autonomous Complex Systems

In Chapter 2, systems of the form $\dot{z} + f(z, \overline{z}) = e^{i\Omega t}$ were considered in which z is a complex variable given in terms of the real variables x and y by $z = x + iy$ and its complex conjugate $\overline{z} = x - iy$ with $i^2 = -1$. Such systems, when converted to autonomous form are three-dimensional. Here we consider four-dimensional autonomous conservative systems of the form

$$\ddot{z} = f(z, \overline{z}). \tag{6.48}$$

Such systems have been studied by Mahmoud and Bountis (2004) among others.

Systems of the form $\ddot{z} = f(z)$ cannot be chaotic, and systems of the form $\ddot{z} = f(\overline{z})$ appear to be unbounded, but eight simple chaotic examples of the form of Eq. (6.48) are listed in Table 6.5 with their corresponding state space plots shown in Fig. 6.28. Model AC_1 is exactly the Hénon–Heiles system in Eq. (6.24) but with x and y interchanged. Model AC_2 is similar to the Hénon–Heiles system but with a sign change, and model AC_3

Table 6.5 Chaotic autonomous complex systems.

Model	Equations	$z_0, \dot{z}_0$	Lyapunov Exponents
AC_1	$\ddot{z} = \bar{z}^2 - z$	$0.5i, 0$	$0.0435, 0, 0, -0.0435$
AC_2	$\ddot{z} = z^2 - \bar{z}$	$0.3 + 0.1i, 0.6i$	$0.0053, 0, 0, -0.0053$
AC_3	$\ddot{z} = \bar{z} - z\bar{z}$	$0.85 + 0.11i, -0.03 - 0.6i$	$0.0039, 0, 0, -0.0039$
AC_4	$\ddot{z} = 1 - z^2\bar{z}$	$-0.4 + 0.2i, -0.3$	$0.0359, 0, 0, -0.0359$
AC_5	$\ddot{z} = z^3 - \bar{z}$	$-0.11 - 0.61i, -0.24 + 1.07i$	$0.0285, 0, 0 - 0.0285$
AC_6	$\ddot{z} = z^2 - z^2\bar{z}$	$0.5 + 0.5i, 0$	$0.0804, 0, 0 - 0.0804$
AC_7	$\ddot{z} = \bar{z}^2 - z^2\bar{z}$	$0.5 + 0.5i, 0$	$0.2479, 0, 0 - 0.2479$
AC_8	$\ddot{z} = z^3 + z\bar{z}$	$-3.79 - 1.72i, 0.89 + 0.37i$	$0.0037, 0, 0, -0.0037$

is a conservative master–slave oscillator that can be written in terms of its real and imaginary parts as

$$\ddot{x} = x - x^2 - y^2$$
$$\ddot{y} = -y. \tag{6.49}$$

Models AC_{4-8} are simple cases with cubic nonlinearities.

Although all eight of these cases are conservative, only three of them have Hamiltonians. As mentioned, AC_1 can be derived from the Hénon–Heiles Hamiltonian in Eq. (6.23) but with x and y (and their respective time derivatives v and u) interchanged. Models AC_4 and AC_7 can be derived from

$$H = \frac{1}{2}\left(v^2 + u^2\right) + \frac{1}{4}\left(x^4 + y^4\right) - \frac{1}{2}x^2y^2 - x \tag{6.50}$$

and

$$H = \frac{1}{2}\left(v^2 + u^2\right) + \frac{1}{4}\left(x^4 + y^4\right) + \frac{1}{2}x^2y^2 - \frac{1}{3}x^3 + xy^2, \tag{6.51}$$

respectively.

6.9 Lotka–Volterra Systems

Chapter 3 included a section on Lotka–Volterra systems, which are popular among population biologists but have much wider application. The system in Eq. (3.28) has chaotic solutions with as few as three species ($N = 3$), but only if some of the a_{ij} parameters are negative, corresponding to predation or cooperation, one example of which was shown in Fig. 3.26.

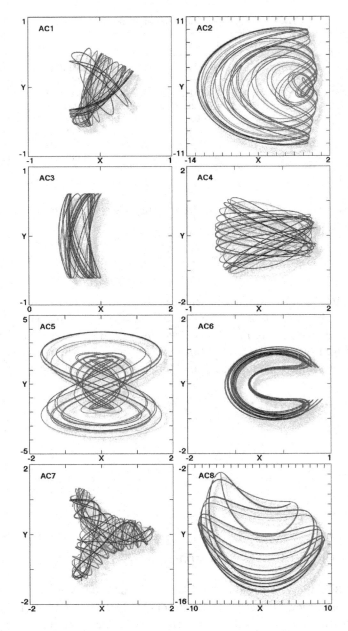

Fig. 6.28   State pace plots for the autonomous complex systems in Table 6.5.

On the other hand, if all the species are competing for a common set of resources (all a_{ij} positive), then the dynamics are restricted to an $(N - 1)$-dimensional *carrying simplex* (Hirsch, 1988), which implies that chaos requires $N \geq 4$. The first such example with $N = 4$ was supplied by Arneodo *et al.* (1982), and a second example was provided by Vano *et al.* (2006). The former was a reformulation of a three-dimensional system with cooperation (Arneodo *et al.*, 1980), and the latter involved an attempt to maximize the Lyapunov exponent, with the consequence that neither system is especially elegant.

A more elegant example with many of the coefficients equal to zero or one is given by

$$
\begin{aligned}
\dot{x} &= x(1 - x - 2y - 1.3z) \\
\dot{y} &= y(1 - y - z - w) \\
\dot{z} &= z(1 - 2x - z) \\
\dot{w} &= w(1 - x - w).
\end{aligned}
\tag{6.52}
$$

This is a dissipative system with an attractor projected onto the xy-plane as shown in Fig. 6.29. As is usual for these cases, one or more species repeatedly comes close to extinction ($y = 0$ in this case) before eventually recovering. Such behavior is not uncommon in ecology.

6.10 Artificial Neural Networks

A popular dynamical model inspired by the connectivity and behavior of neurons in the brain is an *artificial neural network* (Haykin, 1999), one example of which is (Sprott, 2008)

$$
\dot{x}_i = \tanh \sum_{j=1}^{N} a_{ij} x_j - b_i x_i
\tag{6.53}
$$

where N is the number of neurons, each of which represents a dimension of the system. The hyperbolic tangent, also called a *sigmoid function* as shown in Fig. 6.30, is an appropriate nonlinearity because it models the common situation in nature where a small stimulus produces a linear response but the response saturates when the stimulus is large, thereby avoiding unbounded and hence nonphysical solutions. With an appropriate choice of the vector b_i and the matrix a_{ij}, this system can exhibit a wide range of dynamics including chaos, and with a sufficiently large N, it can approximate

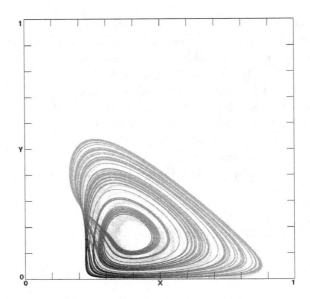

Fig. 6.29   Attractor for the 4-D Lotka–Volterra system in Eq. (6.52) with $(x_0, y_0, z_0, w_0) = (0.5, 0.2, 0.1, 0.5)$, $\lambda = (0.0159, 0, -0.4022, -1.0000)$.

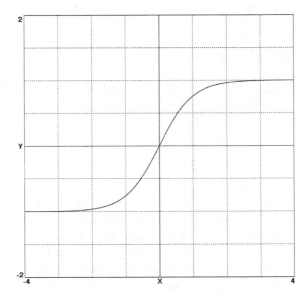

Fig. 6.30   The hyperbolic tangent sigmoid function $y = \tanh x$ is linear for small $|x|$ and saturates at ± 1 for large $|x|$.

to arbitrary accuracy any dynamical system (Funahashi and Nakamura, 1993). This system is equivalent to an alternate form (Sompolinsky *et al.*, 1988) in which the hyperbolic tangent is inside the summation.

As a model of the brain, the signals could represent neural firing rates, and the connections would represent synapses. In a food web, the signals could represent the population of the various species, and the connections would represent feeding. In a financial market, the signals could represent wealth of the investors, and the connections would represent trades. In a political system, the signals could represent voters' position along some political spectrum such as Democrat/Republican, and the connections would be the flow of information between individuals that determine their political views. In general, the network could represent any collection of nonlinearly interacting agents such as people, firms, animals, cells, molecules, or any number of other entities, with the connections representing the flow of energy, data, goods, or information among them. In each case, there is an implicit external source of energy, money, information, or other resource and usually some loss of that resource from the system.

The system has a static equilibrium with all $x_i = 0$ that would represent the state of the system in the absence of external resources (typically death), but it can be driven away from that equilibrium by the positive feedback among the neurons, which implies an external source of energy or other resource not explicit in the equations. The damping coefficients b_i determine the rate at which the system decays to its equilibrium state in the absence of external resources.

6.10.1   *Minimal dissipative artificial neural network*

The minimal dissipative artificial neural network that exhibits chaos (Sprott, 2008) has $N = 4$ and is given by

$$\begin{aligned}
\dot{x} &= \tanh(w - y) - bx \\
\dot{y} &= \tanh(x + w) - by \\
\dot{z} &= \tanh(x + y - w) - bz \\
\dot{w} &= \tanh(z - y) - bw
\end{aligned} \tag{6.54}$$

with its maximum Lyapunov exponent at $b = 0.043$ and an attractor as shown in Fig. 6.31(a).

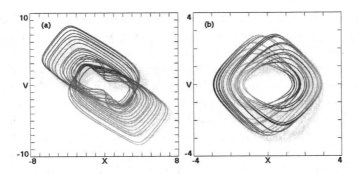

Fig. 6.31   State space plots for artificial neural networks (a) from Eq. (6.54) with $b = 0.043$ and $(x_0, y_0, z_0, w_0) = (2, 1, 1.5, 0)$, $\lambda = (0.0316, 0, -0.0731, -0.1305)$, (b) from Eq. (6.55) with $(x_0, y_0, z_0, w_0) = (0.1, 2.1, -0.8, 0.2)$, $\lambda = (0.0036, 0, 0, -0.0048)$.

6.10.2   *Minimal conservative artificial neural network*

An even simpler conservative artificial neural network with $b = 0$ and $N = 4$ that exhibits chaos is

$$
\begin{aligned}
\dot{x} &= \tanh(y) \\
\dot{y} &= \tanh(z - x) \\
\dot{z} &= \tanh(w) \\
\dot{w} &= \tanh(-z)
\end{aligned}
\tag{6.55}
$$

with a state space plot as shown in Fig. 6.31(b). This case is a master–slave oscillator as evidenced by the fact the last two equations are independent of x and y and produce a periodic nonlinear oscillation whose z component forces the $\dot{y}$ equation that would otherwise combine with the $\dot{x}$ equation to form a second nonlinear oscillator.

6.10.3   *Minimal circulant artificial neural network*

The simplest circulant artificial neural network that exhibits chaos (Sprott, 2008) appears to have $N = 5$ and is given by

$$
\dot{x}_i = -\tanh(x_{i-1} + x_{i+2}) - bx_i
\tag{6.56}
$$

with an attractor for $b = 0.12$ as shown in Fig. 6.32. Other circulant systems, some of which are chaotic for $N < 5$ will be considered in the following chapter.

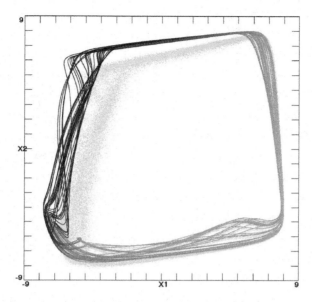

Fig. 6.32 Attractor for the circulant artificial neural network in Eq. (6.56) with $N = 5$, $b = 0.12$, and $x_{i0} = (2, -7, -6, 1, 7.5)$, $\lambda = (0.0248, 0, -0.0827, -0.1042, -0.4379)$.

Chapter 7

Circulant Systems

Systems with dimension greater than four begin to lose their elegance unless they possess some kind of symmetry that reduces the number of parameters. One such symmetry has the variables arranged in a ring of many identical elements, each connected to its neighbors in an identical fashion. The symmetry of the equations is often broken in the solutions, giving rise to spatiotemporal chaotic patterns that are elegant in their own right.

7.1 Lorenz–Emanuel System

One simple circulant ring system was suggested by Lorenz and Emanuel (1998) as a very crude model of some weather variable x_i reported by a succession of weather stations ($1 \leq i \leq N$) spaced equally around the globe along a line of constant latitude. Their model is given by

$$\dot{x}_i = (x_{i+1} - x_{i-2})x_{i-1} - x_i + F \qquad (7.1)$$

in which the ring is closed by assuming $x_{N+1} = x_1, x_0 = x_N$, and $x_{-1} = x_{N-1}$. The nonlinear terms conserve energy and simulate advection, the linear term $-x_i$ represents dissipation of mechanical or thermal energy, and the constant external forcing term F prevents the energy from decaying to zero. The authors studied the case of $F = 8$ and $N = 40$, whose solutions are chaotic.

Such high-dimensional systems raise the question of how best to display the solutions. The state space trajectory can be plotted in the plane of two of the variables as was done for the low-dimensional systems in the previous chapters. Such a plot in the x_1x_2-plane for Eq. (7.1) with $F = 5$ and $N = 101$ is shown in Fig. 7.1, but much information about the remaining

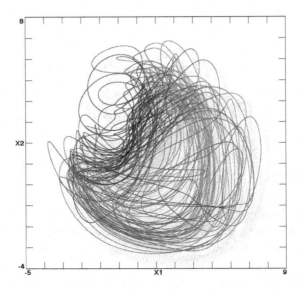

Fig. 7.1   Attractor for the dissipative Lorenz–Emanuel system in Eq. (7.1) projected onto the $x_1 x_2$-plane with $F = 5, N = 101$ ($N_{\min} = 7$), and initial conditions $x_{i0} = \sin(2\pi i/N), \lambda = 0.5392$.

99 dimensions is lost since the attractor may have a dimension on the order of $N/2$. The complexity of the attractor attests to its high dimensionality.

The value of $N = 101$ was chosen to be large but computationally tractable and to be a prime number so that the system cannot break up into a smaller number of subsystems. Generally, the results are qualitatively the same for values of N that exceed some critical number. For Eq. (7.1) with $F = 5$, chaotic solutions are obtained for values of N as small as 7, although it does not follow that all values $N \geq 7$ are chaotic since there may be periodic windows in the range of $7 < N < 101$. The captions of the figures in this chapter will include a value of $N_{\min}$ that represents the minimum value of N for which chaos is obtained for the given parameters and initial conditions as evidenced by a Lyapunov exponent larger than 0.001. In some cases it may be possible to reduce $N_{\min}$ further by altering the parameters and/or the initial conditions.

Another way to display the result is in a spatiotemporal plot as shown in Fig. 7.2 in which the vertical axis represents the location around the ring (the value of i) and the horizontal axis is time. The values of x_i at successive times are plotted in a gray scale in which black is the smallest value of x_i and white is the largest value. Since x_i is defined only at integer values of

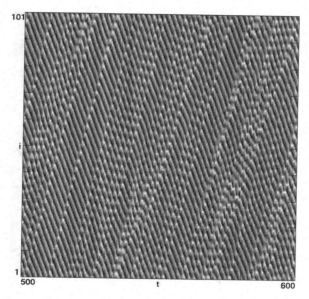

Fig. 7.2 Spatiotemporal plot for the dissipative Lorenz–Emanuel system in Eq. (7.1) with $F = 5, N = 101$ ($N_{\min} = 7$), and initial conditions $x_{i0} = \sin(2\pi i/N)$, $\lambda = 0.5392$.

i, intermediate values (of which there are 7 between each successive pair of integers) are linearly interpolated at each increment of time to determine the gray level for each point in the plot.

Since there is an attractor for the system, the initial conditions are not critical except that they cannot have the same value for all x_{i0} and are here taken as $x_{i0} = \sin(2\pi i/N)$, the single longest-wavelength spatial mode with unit amplitude. The calculation runs for 500 time units before the next 100 time units are plotted to ensure that the system is on the attractor.

Sometimes it is informative to observe the initial transient that occurs while the system is approaching the attractor to verify that the results are independent of the initial conditions and to see how long it takes for the transient to decay. Figure 7.3 shows the transient for the Lorenz–Emanuel system for four different initial conditions. Figure 7.3(a) has the same long wavelength initial conditions as in Fig. 7.2. Figure 7.3(b) has a much shorter initial wavelength corresponding to 32 full waves around the ring. Figure 7.3(c) has all $x_{i0} = 0$ except for $i = 51$ which is $x_{i0} = 1$, and Fig. 7.3(d) has the x_{i0} values taken from a random Gaussian distribution with mean zero and variance one. All cases appear to approach the same attractor in a similar time.

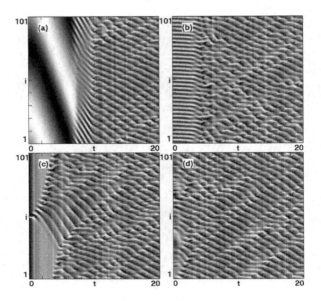

Fig. 7.3   Initial transient for the dissipative Lorenz–Emanuel system in Eq. (7.1) with $F = 5$ and $N = 101$ ($N_{\text{min}} = 7$) for different initial conditions: (a) $x_{i0} = \sin(2\pi i/N)$, (b) $x_{i0} = \sin(64\pi i/N)$, (c) $x_{i0} = 0$ for $i \neq 51$ and $x_{51} = 1$, (d) random Gaussian.

The Lorenz–Emanuel system is also chaotic in the conservative limit given by

$$\dot{x}_i = (x_{i+1} - x_{i-2})x_{i-1} \qquad (7.2)$$

with a spatiotemporal plot for $N = 101$ as shown in Fig. 7.4. Although there is no attractor for such a conservative system, some time is required for the energy to distribute into the spectrum of spatial modes, and this processes is hastened by using an initial condition with a relatively short wavelength (32 full waves around the ring).

Note that we have now introduced yet another type of dimension, the dimension of the network architecture (1 for the case of a ring), to accompany the dimension of the state space ($N = 101$) and the attractor dimension. A one-dimensional ring network is often assumed to have only interactions between nearest neighbors around the ring, but we relax that definition slightly to allow connections with other elements around the ring as long as the connectivity is uniform around the ring as in the examples in this chapter. In such a case, the matrix of connectivities is said to be *circulant*.

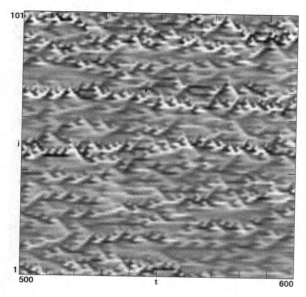

Fig. 7.4 Spatiotemporal plot for the conservative Lorenz–Emanuel system in Eq. (7.2) with $N = 101$ ($N_{min} = 5$), and initial conditions $x_{i0} = \sin(64\pi i/N)$, $\lambda = 0.5274$.

Many other architectures are possible including two-dimensional *grids* such as might be used to describe landscapes or the weather, and three-dimensional *lattices* that might describe crystals or atmospheric motion or perhaps the connection of neurons in the brain. Such higher-dimensional spatial networks will not be considered here because of their large computational demands, but they represent a fruitful area of research as computers become ever more powerful.

7.2 Lotka–Volterra Systems

Lotka–Volterra systems were described in Chapter 3, with the standard form given by Eq. (3.28). Such models are popular among population biologists but have application to any situation in which a number of agents are competing for fixed, limited resources. While perhaps not very biologically realistic, simple high-dimensional extensions are possible in which the agents are arranged in a ring with each agent competing only with a small number of its near neighbors and with the interactions uniform around the ring. A situation where such a model with one spatial dimension might be useful is to explain the ecology in a long, narrow, slowly flowing river.

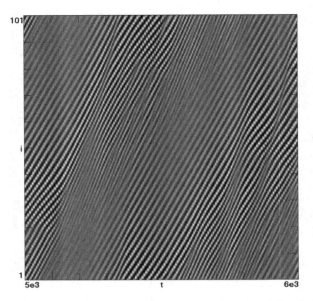

Fig. 7.5   Spatiotemporal plot for the Lotka–Volterra ring system in Eq. (7.3) with $N =$ 101 ($N_{min} = 59$) and initial conditions $x_{i0} = 0.4 + 0.3\sin(2\pi i/N)$, $\lambda = 0.0078$.

One of the simplest such models was proposed by Sprott *et al.* (2005b) and given by

$$\dot{x}_i = x_i(1 - x_{i-2} - x_i - x_{i+1}). \tag{7.3}$$

It seems peculiar for each agent to compete with the one on its right (x_{i+1}) and the second on its left (x_{i-2}) as well as itself (x_i) but to ignore the one on its immediate left (x_{i-1}), but perhaps that one is a family member. In any event, the model exhibits interesting chaotic dynamics as shown in the spatiotemporal plot in Fig. 7.5 for $N = 101$. Notice the short-wavelength, high-frequency oscillation, with each site out of phase with its nearest neighbors, and propagating waves superimposed on what might be regarded as weak turbulence if it were to occur in a fluid.

This system is similar to the Lorenz–Emanuel system in the sense that it is quadratic, dissipative, circulant, and involves interaction of each agent with itself and only two or three near neighbors, but it is slightly more complicated since the forcing is linear (x_i) while the dissipation is nonlinear $(-x_i^2)$, in contrast to the Lorenz–Emanuel system in which both the forcing and the dissipation are linear. Furthermore, the variables are limited to positive (or negative) values, depending on the initial conditions, and if any variable ever goes to zero, it will remain there forever no matter what its neighbors do. In a biological system, this would correspond to extinction of a species, or in a financial system to a firm or individual going bankrupt. In such a case, the ring would break into a number of isolated segments with fixed boundary conditions, and the chaos is likely to be suppressed.

7.3 Antisymmetric Quadratic System

The previous case lacks symmetry in its neighborhood. In fact, it can be shown that a symmetric Lotka–Volterra ring system cannot exhibit chaos (Wildenberg *et al.*, 2005). However, there are antisymmetric systems that involve quadratic interactions with only the two nearest neighbors on each side, one of the simplest examples of which is

$$\dot{x}_i = 8x_i(x_{i+1} - x_{i-1}) - x_i(x_{i+2} - x_{i-2}). \qquad (7.4)$$

This is a conservative system that exhibits chaos with a spatiotemporal plot as shown in Fig. 7.6, and it has the biologically plausible feature that the interaction between the nearest neighbors is considerably larger than the interaction between the second nearest neighbors.

7.4 Quadratic Ring System

The preceding examples contain quadratic nonlinearities but require interactions with neighbors that are not adjacent. In a true ring system, the interactions would only involve the nearest neighbors. Perhaps the simplest such case, with some resemblance to the Lotka–Volterra system, but with only two nonlinearities, is given by

$$\dot{x}_i = x_{i+1}(1 - x_{i+1}) - x_{i-1}(1 - x_{i-1}). \qquad (7.5)$$

This is a conservative system that is weakly chaotic when the initial condition has a sufficiently large amplitude with a spatiotemporal plot as shown in Fig. 7.7.

7.5 Cubic Ring System

If we allow cubic nonlinearities, there is an especially elegant case (Brummitt and Sprott, 2009) that only involves nearest neighbor interactions

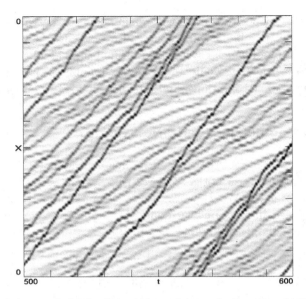

Fig. 7.6 Spatiotemporal plot for the antisymmetric quadratic system in Eq. (7.4) with $N = 101$ ($N_{\min} = 7$) and initial conditions $x_{i0} = -0.1 + 0.1\sin(2\pi i/N), \lambda = 0.0842$.

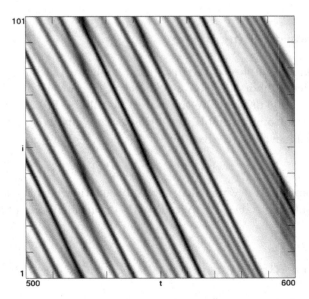

Fig. 7.7 Spatiotemporal plot for the quadratic ring system in Eq. (7.5) with $N = 101$ ($N_{\min} = 101$) and initial conditions $x_{i0} = 0.15\sin(2\pi i/N), \lambda = 0.0010$.

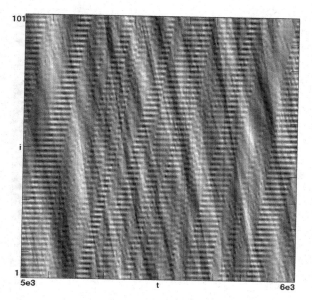

Fig. 7.8 Spatiotemporal plot for the cubic ring system in Eq. (7.6) with $N = 101$ ($N_{\min} = 13$) and initial conditions $x_{i0} = 4\sin(2\pi i/N), \lambda = 0.0614$.

given by

$$\dot{x}_i = (x_{i+1} - x_{i-1})^3. \tag{7.6}$$

This is a conservative system in which the right-hand side can be thought of as proportional to the cube of the spatial gradient of x around the ring. Its spatiotemporal plot as shown in Fig. 7.8 is chaotic when the initial condition has a sufficiently large amplitude, but most of the energy resides in the shortest and longest wavelength modes as is evident from the figure which resembles wrinkled corduroy. In the corresponding partial differential equation, the energy continues to pile up at progressively shorter wavelengths in a kind of 'ultraviolet catastrophe.' The Lyapunov exponent scales as the square of the amplitude of the initial condition as can be seen from the fact that x^2 has units of inverse time, and hence Eq. (7.6) can be linearly rescaled.

7.6 Hyperlabyrinth System

The labyrinth system proposed by Thomas (1999) and given by Eq. (4.9) is easily generalized to arbitrary dimension (Thomas *et al.*, 2004) where it

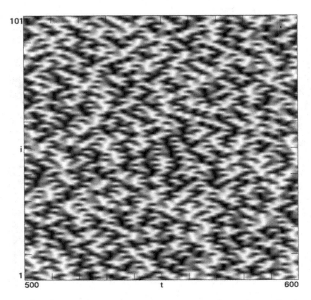

Fig. 7.9   Spatiotemporal plot for the hyperlabyrinth system in Eq. (7.7) with $N = 101$ ($N_{\min} = 3$) and initial conditions $x_{i0} = \sin(2\pi i/N)$, $\lambda = 0.4188$.

takes the particularly simple form

$$\dot{x}_i = \sin x_{i+1}. \qquad (7.7)$$

This conservative system and its dissipative generalization were studied in some detail by Chlouverakis and Sprott (2007) who called it 'hyperlabyrinth chaos' since it can be considered as the trajectory of a particle that executes a chaotic walk through an N-dimensional labyrinth (or on the surface of an N-torus), analogous to the three-dimensional case in Fig. 4.11. However, it can equally well be treated as a symmetric ring of agents, each responding only to the one on its immediate right (or left) around the ring. Its spatiotemporal plot is shown in Fig. 7.9, where the gray level in this case is proportional to $\sin x_i$ rather than to x_i since the trajectory can wander to arbitrarily large values of $|x_i|$. It was already shown in Chapter 4 that this system is chaotic even for $N = 3$.

7.7   Circulant Neural Networks

The previous chapter discussed artificial neural networks, which can also be circulant, in which case the matrix of connections a_{ij} becomes a vector

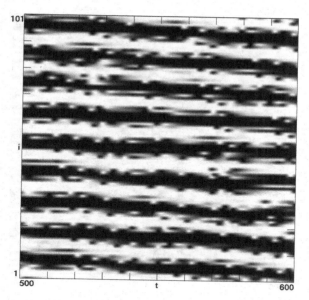

Fig. 7.10 Spatiotemporal plot for the circulant neural network in Eq. (7.9) with $b = 0.37, N = 101$ ($N_{\min} = 8$), and initial conditions $x_{i0} = 0.1 \sin(18\pi i/N), \lambda = 0.0325$.

a_j, and the corresponding equations become

$$\dot{x}_i = \tanh \sum_{j=1}^{N-1} a_j x_{i+j} - bx_i. \qquad (7.8)$$

A large circulant network of this type with localized connections is given by

$$\dot{x}_i = \tanh(x_{i+1} + x_{i+2} - x_{i+3} - x_{i+4} + x_{i+5} - x_{i+6}) - bx_i, \qquad (7.9)$$

which is chaotic with $b = 0.5$ for $N = 317$ (Sprott, 2008). However, with $N = 101$, a smaller value of b is required. Figure 7.10 shows the spatiotemporal plot for this case with $b = 0.37$.

It turns out that there is an even simpler such network with the same size neighborhood but centered on the site and with only three connections given by

$$\dot{x}_i = \tanh(x_{i+3} - x_{i+1} - x_{i-3}) - bx_i, \qquad (7.10)$$

which is chaotic with $b = 0.1$ for $N = 101$. The spatiotemporal plot for this case is shown in Fig. 7.11.

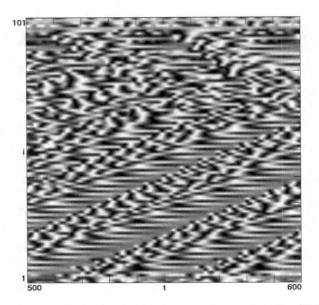

Fig. 7.11   Spatiotemporal plot for the circulant neural network in Eq. (7.10) with $b = 0.1, N = 101$ ($N_{\min} = 13$), and initial conditions $x_{i0} = \sin(2\pi i/N), \lambda = 0.0253$.

7.8   Hyperviscous Ring

A somewhat less elegant example of a dissipative ring system with a quadratic nonlinearity representing the simplest spatially discrete approximation to the Kuramoto–Sivashinsky partial differential equation (Kuramoto and Tsuzuki, 1976; Sivashinsky, 1977) discussed in the next chapter is given by

$$\dot{x}_i = x_i(x_{i+1} - x_{i-1}) - 5x_i + 3.5(x_{i+1} + x_{i-1}) - (x_{i+2} + x_{i-2}) \quad (7.11)$$

with a spatiotemporal plot as shown in Fig. 7.12.

7.9   Rings of Oscillators

In the previous examples, each of the N nodes in the ring is described by a first-order ODE in a single variable x_i. The nodes, of course, can have a more complicated description in terms of higher-order differential equations or with multiple variables. For example, the nodes could be oscillators coupled to their neighbors in a similar way to the cases with two

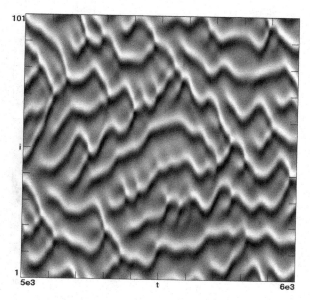

Fig. 7.12 Spatiotemporal plot for the hyperviscous ring system in Eq. (7.11) with $N = 101$ ($N_{\min} = 18$) and initial conditions $x_{i0} = \sin(2\pi i/N)$, $\lambda = 0.0250$.

coupled oscillators in the previous chapter. It is not difficult to find chaotic examples of such systems.

7.9.1 *Coupled pendulums*

A conceptually simple example of such a system is a ring of identical frictionless pendulums each coupled to its two nearest neighbors by a torsional spring in the manner described by Eq. (6.8). When generalized to a ring of arbitrary size, the governing equations can be written simply as

$$\ddot{x}_i + \sin x_i = k(x_{i-1} - 2x_i + x_{i+1}) \qquad (7.12)$$

where k is the spring constant for the springs, which are assumed to be identical. The spatiotemporal plot for this conservative case with $k = 1$ is shown in Fig. 7.13.

7.9.2 *Coupled cubic oscillators*

All of the other nonlinear oscillators in the previous chapter can be connected in a ring with a variety of couplings, both linear and nonlinear, and

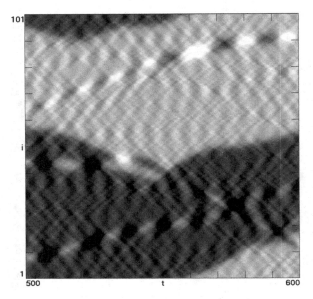

Fig. 7.13 Spatiotemporal plot for the ring of coupled frictionless pendulums in Eq. (7.12) with $k = 1, N = 101$ ($N_{min} = 5$), and initial conditions $x_{i0} = 4\sin(8\pi i/N), v_{i0} = 0, \lambda = 0.0919$.

it is easy to find parameters for them that produce chaos. A simple example is a ring of identical conservative oscillators with a cubic restoring force, each driven linearly by the one on its right around the ring and governed by the equation

$$\ddot{x}_i + x_i^3 = kx_{i+1}. \tag{7.13}$$

This mater–slave system has the peculiar property that each oscillator is a slave to the one on its right but is a master to the one on its left. Such behavior would be unusual for a mechanical system constrained by Newton's third law (action and reaction), and even more so for a social system, but it might represent a military chain of command leading up to the President, who is elected by the privates, or a food chain in which the smallest virus feeds on its human host. The spatiotemporal plot for this system with $k = 0.1$ is shown in Fig. 7.14.

7.9.3   *Coupled signum oscillators*

An example of a ring of identical conservative oscillators with a signum nonlinearity, each driven linearly by the one on its right around the ring, is

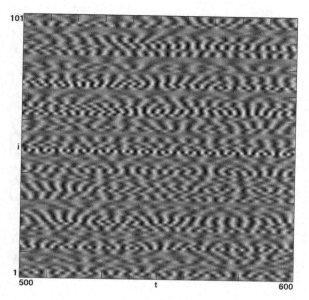

Fig. 7.14 Spatiotemporal plot for the ring of coupled conservative cubic oscillators in Eq. (7.13) with $k = 0.1, N = 101$ ($N_{min} = 3$), and initial conditions $x_{i0} = \sin(8\pi i/N), v_{i0} = 0, \lambda = 0.0401$.

given by the equation

$$\ddot{x}_i + \mathrm{sgn}\, x_i = kx_{i+1}. \tag{7.14}$$

The spatiotemporal plot for this system with $k = 0.1$ is shown in Fig. 7.15. The signum nonlinearity is especially appropriate for electronic implementation, and an electrical circuit model for a system similar to this is shown in Chapter 10. As usual, the $\mathrm{sgn}\, x_i$ term is replaced with $\tanh(500x_i)$ for the purpose of establishing that the chaos is not a numerical artifact and for estimating the Lyapunov exponent.

7.9.4 *Coupled van der Pol oscillators*

The previous examples of coupled oscillators involved conservative systems in which the nonlinearity is in the restoring force term. It is also possible to couple dissipative oscillators and oscillators having other types of nonlinearities. There are countless such combinations, but we consider here a particularly simple and familiar system, a ring of coupled van der Pol oscillators given by

$$\ddot{x}_i + b(x_i^2 - 1)\dot{x}_i + x_i = k_1\dot{x}_{i+1} - k_2 x_{i+1}. \tag{7.15}$$

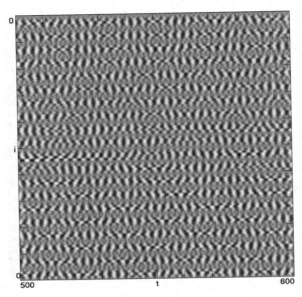

Fig. 7.15 Spatiotemporal plot for the ring of coupled conservative signum oscillators in Eq. (7.14) with $k = 0.1, N = 101$ ($N_{min} = 7$), and initial conditions $x_{i0} = 0.3\sin(20\pi i/N)$, $v_{i0} = 0, \lambda = 0.0324$.

This system is a ring of master–slave limit cycle oscillators with each oscillator driven by the one on its right. Although most of the research on coupled van der Pol oscillators has focused on the synchronization of the limit cycles (Woafo and Kadki, 2004), such systems can also exhibit chaos. In particular, Eq. (7.15) is chaotic with a spatiotemporal plot as shown in Fig. 7.16.

7.9.5 *Coupled FitzHugh–Nagumo oscillators*

In Eq. (6.10) an example was given of two coupled FitzHugh–Nagumo oscillators. Such oscillators can also be coupled in a ring, with a simple example given by

$$\dot{x}_i = x_i - x_i^3 - y_i + kx_{i+1} \qquad (7.16)$$
$$\dot{y}_i = a + bx_i - cy_i.$$

As with the previous example of coupled van der Pol oscillators, this system is a ring of master–slave limit cycle oscillators with each oscillator driven by the one on its right. Its spatiotemporal plot is shown in Fig. 7.17.

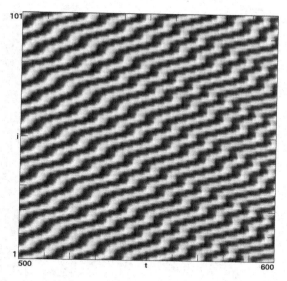

Fig. 7.16 Spatiotemporal plot for the ring of coupled van der Pol oscillators in Eq. (7.15) with $(b, k_1, k_2, N) = (0.5, 3, 1, 101)$ ($N_{\min} = 13$) and initial conditions $x_{i0} = \sin(32\pi i/N), v_{i0} = 0, \lambda = 0.0236$.

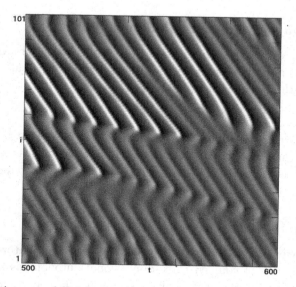

Fig. 7.17 Spatiotemporal plot for the ring of coupled FitzHugh–Nagumo oscillators in Eq. (7.16) with $(a, b, c, k, N) = (0.6, 1, 0.1, 0.58, 101)$ ($N_{\min} = 21$) and initial conditions $x_{i0} = \sin(16\pi i/N), y_{i0} = 0, \lambda = 0.0258$.

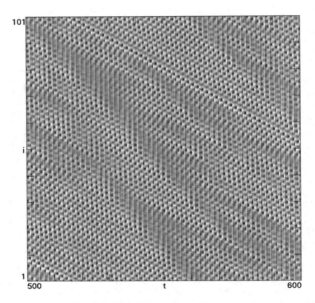

Fig. 7.18 Spatiotemporal plot for the ring of coupled complex oscillators in Eq. (7.17) with $(a, k, N) = (0.7, 1, 101)$ ($N_{\min} = 17$) and initial condition $z_{i0} = -0.7\sqrt{-1}\cos(101\pi i/N), \lambda = 0.0445$.

7.9.6 Coupled complex oscillators

In Eq. (6.11) an example was given of two coupled oscillators with complex variables and quadratic nonlinearities. Such oscillators can also be coupled in a ring, with a simple example given by

$$\dot{z}_i + z_i^2 - a\overline{z}_i + 1 = -kz_{i+1} \qquad (7.17)$$

whose spatiotemporal plot is shown in Fig. 7.18. Note that adjacent sites tend to alternate and that in the initial condition, which is a bit critical, the symbol i is the position around the ring and not $\sqrt{-1}$, which is written explicitly.

7.9.7 Coupled Lorenz systems

A logical extension of the two-dimensional coupled oscillators considered above is the case of coupled three-dimensional autonomous systems that are individually chaotic. Any of the systems in Chapters 3 and 4 are candidates for such a system. In general, when the individual oscillators are chaotic, it is simple to obtain chaos in a ring of such oscillators with sufficiently weak coupling. When the coupling is very strong, there is a tendency for the oscillators to synchronize, in which case they behave like a single oscillator, which may be chaotic but lacks spatial structure (Afraimovich and Lin, 1998; Chiu *et al.*, 2001). The more interesting examples are those with intermediate coupling where spatiotemporal chaotic patterns with discernable structure develop.

One of the most widely studied example of such a case involves coupled Lorenz systems. Such systems are especially good because the basin of attraction for each Lorenz system is the entire state space, and so they tend to be bounded when coupled in a ring, although boundedness is not guaranteed.

7.9.7.1 *Viscously coupled case*

Many different couplings are possible, but we consider here a case with viscous coupling proposed by Jackson and Kodogeorgiou (1992) as a model of Rayleigh–Bénard turbulence in which adjacent fluid cells do not necessarily counter-rotate. The resulting system is

$$\begin{aligned}
\dot{x}_i &= \sigma(y_i - x_i) - k(x_{i+1} + 2x_i + x_{i-1}) \\
\dot{y}_i &= -x_i z_i + r x_i - y_i \\
\dot{z}_i &= x_i y_i - b z_i,
\end{aligned} \tag{7.18}$$

which for the standard parameters used by Lorenz (1963) and a coupling of $k = 1$, produces the spatiotemporal plot in Fig. 7.19. The plot tends to have either black or white regions, corresponding to which lobe each attractor is on, with the limited gray regions corresponding to the relatively abrupt transitions from one lobe to the other. There is a tendency for adjacent sites to alternate in sign, corresponding to counter-rotation, but with many exceptions.

7.9.7.2 *Diffusively coupled case*

The case with diffusive coupling has been studied by Josić and Wayne (2000) and is given in its simplest form by

$$\begin{aligned}
\dot{x}_i &= \sigma(y_i - x_i) + k(x_{i+1} - 2x_i + x_{i-1}) \\
\dot{y}_i &= -x_i z_i + r x_i - y_i \\
\dot{z}_i &= x_i y_i - b z_i
\end{aligned} \tag{7.19}$$

with a spatiotemporal plot as shown in Fig. 7.20. The coupling term in this case is a discrete approximation to the Laplacian $\nabla^2 x$ and tends to synchronize adjacent sites (with k positive), leading to large-scale spatiotemporal

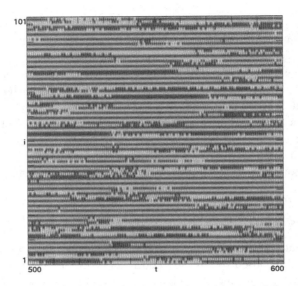

Fig. 7.19   Spatiotemporal plot for the ring of viscously coupled Lorenz systems in Eq. (7.18) with $(\sigma, r, b, k, N) = (10, 28, 8/3, 1, 101)$ $(N_{\min} = 2)$ and initial conditions $(x_{i0}, y_{i0}, z_{i0}) = (10\sin(100\pi i/N), 0, 0), \lambda = 0.8829$.

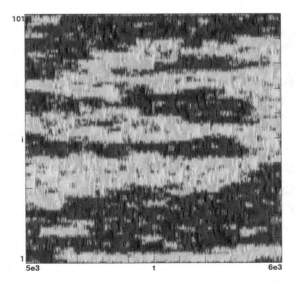

Fig. 7.20   Spatiotemporal plot for the ring of diffusively coupled Lorenz systems in Eq. (7.19) with $(\sigma, r, b, k, N) = (10, 28, 8/3, 1, 101)$ $(N_{\min} = 1)$ and initial conditions $(x_{i0}, y_{i0}, z_{i0}) = (10\sin(8\pi i/N), 0, 0), \lambda = 0.08835$.

chaotic structures. Again, the regions tend to be either black or white corresponding to which lobe each attractor is on, with relatively small regions of gray. Recall that the linear diffusion equation with constant diffusion coefficient D (also called the *heat equation*) is given by $\partial u/\partial t = \nabla^2 u$, and so Eq. (7.19) can be regarded as a heat equation coupled to a collection of Lorenz attractors. Note that, unlike all the previous examples, this case is chaotic even for $N = 1$ since the coupling term is then identically zero and the system reduces to a single chaotic Lorenz system.

7.9.7.3 *Coupled diffusionless case*

Although these examples were motivated by familiarity and applicability to physical systems, they are not the most elegant examples of coupled Lorenz systems. For that purpose, we take instead the diffusionless Lorenz system in Eq. (3.2) and simplify the coupling so that it is unidirectional as given by

$$\dot{x}_i = y_i - x_i + kx_{i+1}$$
$$\dot{y}_i = -x_i z_i \qquad (7.20)$$
$$\dot{z}_i = x_i y_i - 1,$$

which preserves the spirit of the previous cases and produces a spatiotemporal plot for $k = 0.5$ and $N = 101$ as shown in Fig. 7.21. The plot in many ways more nearly resembles fluid turbulence than does Fig. 7.19 or Fig. 7.20.

7.9.8 *Coupled jerk systems*

As a final example of a ring of coupled chaotic oscillators, consider the simplest quadratic chaotic system in Eq. (3.8) coupled in a ring as follows:

$$\dddot{x}_i + a\ddot{x}_i - \dot{x}_i^2 + x_i = -kx_{i+1}. \qquad (7.21)$$

This system is rather delicate because of the narrow range of a over which chaos occurs, but it exhibits spatiotemporal chaos for $a = 2.12$ and $k = 0.006$ with a spatiotemporal plot as shown in Fig. 7.22.

7.10 Star Systems

The ring architecture is not the only possible circulant system, although it is the most studied. An alternate configuration is the *star* in which each

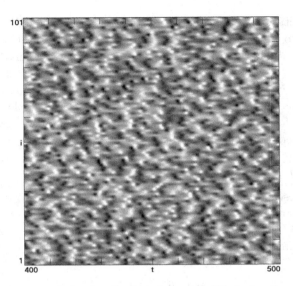

Fig. 7.21 Spatiotemporal plot for the ring of coupled diffusionless Lorenz systems in Eq. (7.20) with $k = 0.5, N = 101$ ($N_{min} = 3$), and initial conditions $(x_{i0}, y_{i0}, z_{i0}) = (2\sin(24\pi i/N), 0, 0), \lambda = 0.2642$.

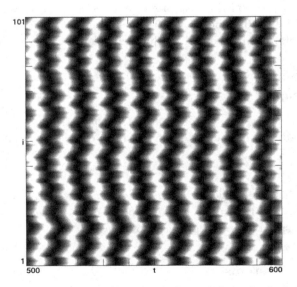

Fig. 7.22 Spatiotemporal plot for the ring of coupled quadratic jerk systems in Eq. (7.21) with $(a, k, N) = (2.12, 0.006, 101)$ ($N_{min} = 5$) and initial conditions $(x_{i0}, y_{i0}, z_{i0}) = (4 + \sin(64\pi i/N), 2, 0), \lambda = 0.0.0138$.

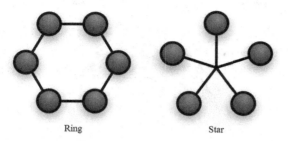

Ring   Star

Fig. 7.23   Two types of circulant networks.

element is symmetrically coupled to all the other elements as shown in Fig. 7.23. In the case of $N = 3$, these cases are called the *delta* (Δ) and the *wye* (Y), respectively, for obvious reasons, especially in electrical power systems. There are countless such possibilities, but we show here only a few examples.

7.10.1   *Coupled pendulums*

Analogous to the ring of pendulums in Eq. (7.12), consider the same N frictionless pendulums arranged in a star and linearly coupled according to

$$\ddot{x}_i + \sin x_i = \frac{1}{N} \sum_{j=1}^{N} x_j - x_i. \tag{7.22}$$

For such a case, all the pendulums interact equally with all the others, and so there is no sense of a spatial neighborhood. This in a simple example of a *globally coupled network* (everything is coupled to everything else with no regard for its spatial location). Consequently, it does not make sense to display the results in a spatiotemporal plot, and so instead we revert to a state space plot as shown in Fig. 7.24 for $N = 101$ in which the position x_1 and velocity v_1 of one typical pendulum is plotted with the third coordinate given by the position x_2 of one of the other pendulums. Eq. (7.22) is an extreme example of a *mean field approximation* (Amari *et al.*, 1977) since the behavior of each pendulum is controlled by the average of all the others.

7.10.2   *Coupled cubic oscillators*

Analogous to the ring of conservative nonlinear oscillators with a cubic restoring force in Eq. (7.13), consider the same N cubic oscillators arranged

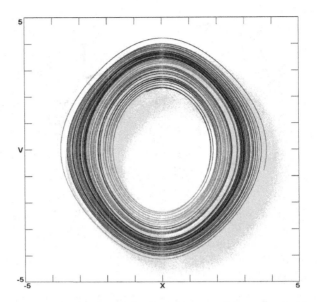

Fig. 7.24 State space plot for one of the frictionless pendulums coupled in a star as in Eq. (7.22) with $N = 101$ ($N_{\min} = 3$) and initial conditions $x_{i0} = 4\sin(4\pi i/N), v_{i0} = 0, \lambda = 0.0369$.

in a star and linearly coupled according to

$$\ddot{x}_i + x_i^3 = \frac{1}{N} \sum_{j=1}^{N} x_j. \tag{7.23}$$

The state space plot for this case is shown in Fig. 7.25 with $N = 101$.

7.10.3 *Coupled signum oscillators*

Analogous to the ring of conservative nonlinear oscillators with a signum restoring force in Eq. (7.14), consider the same N nonlinear oscillators arranged in a star and linearly coupled according to

$$\ddot{x}_i + \operatorname{sgn} x_i = \frac{1}{N} \sum_{j=1}^{N} x_j. \tag{7.24}$$

The state space plot for this case is shown in Fig. 7.26 with $N = 101$.

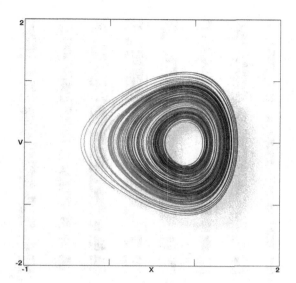

Fig. 7.25   State space plot for one of the conservative cubic oscillators coupled in a star as in Eq. (7.23) with $N = 101$ ($N_{\min} = 3$), and initial conditions $x_{i0} = \sin(4\pi i/N), v_{i0} = 0, \lambda = 0.0319$.

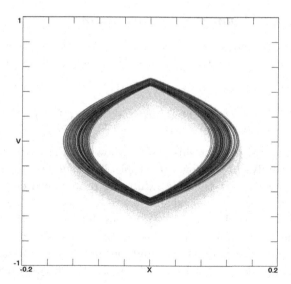

Fig. 7.26   State space plot for one of the conservative signum oscillators coupled in a star as in Eq. (7.24) with $N = 101$ ($N_{\min} = 35$), and initial conditions $x_{i0} = 0.9\sin(4\pi i/N), v_{i0} = 0, \lambda = 0.0659$.

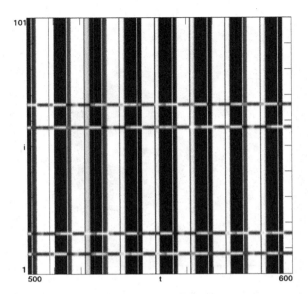

Fig. 7.27 Spatiotemporal plot for the van der Pol oscillators coupled in a star as in Eq. (7.25) with $(b, k, N) = (2.5, 0.45, 101)$ ($N_{min} = 29$) and initial conditions $x_{i0} = \sin(4\pi i/N), v_{i0} = 0, \lambda = 0.0842$.

7.10.4 Coupled van der Pol oscillators

Analogous to the ring of coupled van der Pol oscillators in Eq. (7.15), consider the same N van der Pol oscillators arranged in a star and linearly coupled according to

$$\ddot{x}_i + b(x_i^2 - 1)\dot{x}_i + x_i = \frac{k}{N} \sum_{j=1}^{N} (x_j + \dot{x}_j). \qquad (7.25)$$

The spatiotemporal plot for this case as shown in Fig. 7.27 indicates that the chaos is entirely due to a few sites (typically about 4 or 5), as determined by the initial conditions, that are exhibiting pronounced chaotic behavior while the others are synchronized and very nearly periodic. The background of periodic oscillators behaves like a strong forcing function for the few oscillators that are mainly responsible for the chaos. Of course there is no sense of a neighborhood in this case, and the chaotic sites could be anywhere around the ring, but they tend to retain their identity as determined by the initial conditions.

Figure 7.28 shows two of the oscillators, the one at $i = 1$ that is essentially a limit cycle alongside the one at $i = 8$ that exhibits chaos. The

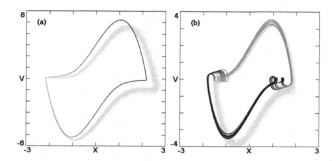

Fig. 7.28 Attractors for two of the van der Pol oscillators coupled in a star as in Eq. (7.25) with $(b, k, N) = (2.5, 0.45, 101)$ $(N_{min} = 29)$ and initial conditions $x_{i0} = \sin(4\pi i/N), v_{i0} = 0$, (a) one nearly periodic, and (b) the other apparently chaotic.

region of parameter space over which chaos occurs is relatively narrow for this system.

7.10.5   *Coupled FitzHugh–Nagumo oscillators*

Analogous to the ring of coupled FitzHugh–Nagumo oscillators in Eq. (7.16), consider the same N coupled FitzHugh–Nagumo oscillators arranged in a star and linearly coupled according to

$$\dot{x}_i = x_i - x_i^3 - y_i - \frac{k}{N} \sum_{j=1}^{N} x_j$$
$$\dot{y}_i = a + bx_i - cy_i \tag{7.26}$$

whose attractor for $N = 101$ is shown in Fig. 7.29.

7.10.6   *Coupled complex oscillators*

Analogous to the ring of coupled oscillators with complex variables and quadratic nonlinearities in Eq. (7.17), consider the same N coupled complex oscillators arranged in a star and linearly coupled according to

$$\dot{z}_i + az_i^2 + b\overline{z}_i + 1 = -\frac{k}{N} \sum_{j=1}^{N} x_j \tag{7.27}$$

whose state space plot for $N = 101$ is shown in Fig. 7.30.

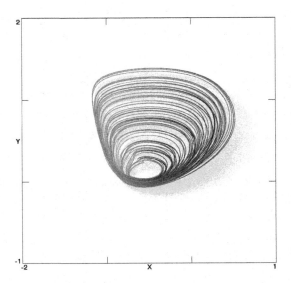

Fig. 7.29 Attractor for one of the FitzHugh–Nagumo oscillators coupled in a star as in Eq. (7.26) with $(a, b, c, k, N) = (0.28, 0.5, 0.04, 1, 101)$ ($N_{min} = 4$) and initial conditions $(x_{i0}, y_{i0}) = (0.1 \sin(16\pi i/N), 0), \lambda = 0.0008$.

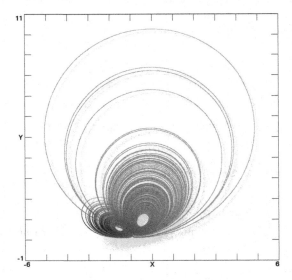

Fig. 7.30 State space plot for one of the complex oscillators coupled in a star as in Eq. (7.27) with $(k, a, b, N) = (1.7, -0.4, 1, 101)$ ($N_{min} = 15$) and initial conditions $z_{i0} = e^{\sqrt{-1}\pi i}, \lambda = 0.0335$.

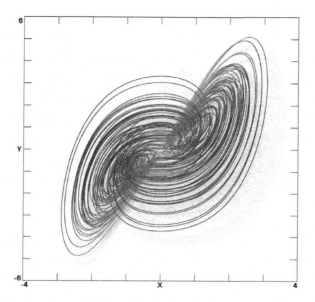

Fig. 7.31 Attractor for one of the diffusionless Lorenz systems coupled in a star as in Eq. (7.28) with $N = 101$ ($N_{\min} = 1$) and initial conditions $(x_{i0}, y_{i0}, z_{i0}) = (2\sin(24\pi i/N), 0, 0)$, $\lambda = 0.2631$.

7.10.7 *Coupled diffusionless Lorenz systems*

Analogous to the ring of coupled diffusionless Lorenz systems in Eq. (7.20), consider the same N diffusionless Lorenz systems arranged in a star and linearly coupled according to

$$\dot{x}_i = y_i - x_i - \sum_{j=1}^{N} x_j$$
$$\dot{y}_i = -x_i z_i \qquad\qquad\qquad (7.28)$$
$$\dot{z}_i = x_i y_i - 1$$

whose attractor for $N = 101$ is shown in Fig. 7.31.

The attractor looks quite similar to the single diffusionless Lorenz attractor in Fig. 3.3 since the sum of all the x values remains close to zero. However, the Lyapunov exponent is somewhat larger than for the case of a single attractor, indicating that the globally coupled system is different and more chaotic. Its spatiotemporal plot (not shown) looks identical to Fig. 7.21 and confirms that the individual Lorenz systems are not synchronized despite the large coupling.

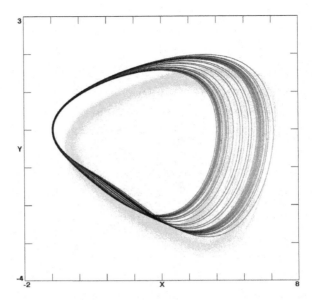

Fig. 7.32 Attractor for one of the quadratic jerk systems coupled in a star as in Eq. (7.29) with $(a, k, N) = (2.06, 0.005, 101)$ ($N_{\min} = 5$) and initial conditions $(x_{i0}, y_{i0}, z_{i0}) = (4 + \sin(64\pi i/N), 2, 0)$, $\lambda = 0.0120$.

7.10.8   *Coupled jerk systems*

Analogous to the ring of coupled jerk systems in Eq. (7.21), consider the same N jerk systems arranged in a star and linearly coupled according to

$$\dddot{x}_i + a\ddot{x}_i - \dot{x}_i^2 + x_i = \frac{k}{N} \sum_{j=1}^{N} x_j. \qquad (7.29)$$

whose attractor for $a = 2.06$, $k = 0.005$ and $N = 101$ is shown in Fig. 7.32. The coupling is very weak for this case as it was for Eq. (7.21), but the oscillators are neither synchronized nor completely independent.

Chapter 8

Spatiotemporal Systems

Many systems in the real world are too complicated to be adequately described by a small number of coupled ordinary differential equations. Important examples include spatiotemporal systems such as the weather or other moving fluids, which are best described by partial differential equations (PDEs). Despite their inherent complexity, such 'infinite-dimensional' chaotic systems can be very mathematically elegant and can produce spatiotemporal chaos.

8.1 Numerical Methods

Most spatiotemporal systems can be approximated by a very large number of ODEs, each representing one spatial location and coupled to its near neighbors. The simplest such examples involve a single spatial dimension and can be thought of as the continuum limit of the ring systems in the previous chapter. With infinitely many ODEs spaced infinitesimally close together around the ring, the resulting dynamical system is *infinite-dimensional* in the sense that infinitely many spatial variables are involved, each requiring an initial condition, and thus the initial condition must be specified by a continuous function of space. There are many systems with two and three spatial dimensions that are important in real-world applications and that exhibit interesting spatiotemporal dynamics, but these go beyond the scope of this book. To avoid confusion, we will change notation from earlier chapters and use the variable u to represent the dynamic quantity that varies in space x and time t, and we will assume that $u(x,t)$ obeys some PDE whose elegance is optimized as with the previous ODEs.

Typically, PDEs that are capable of producing spatiotemporal chaos involve the partial derivatives of u with respect to time $\partial u/\partial t$ and with respect to position $\partial u/\partial x$ and higher derivatives such as $\partial^2 u/\partial t^2$, $\partial^2 u/\partial x^2$, and $\partial^2 u/\partial x \partial t$. To simplify the notation, we will represent the respective derivatives as $u_t, u_x, u_{tt}, u_{xx}, u_{xt}$, and so forth. In systems with two or more spatial dimensions $x, y, z, \ldots$, chaos is relatively common, but it can also occur in systems with a single spatial dimension, although it is relatively rare.

Furthermore, we will be concerned only with examples in which the chaos occurs in both space and time, although purely temporal chaos in which spatial locations are synchronized and purely spatial chaos in which the temporal behavior is constant or periodic are also possible (Pesin and Sinai, 1991). Purely spatial chaos is precluded by the fact that the systems considered here are assumed to lie on a ring with periodic boundary conditions, and thus the only spatial modes that are allowed are those in which an integer number of wavelengths fit around the ring. Other boundary conditions are possible and common, but they will not be considered here since the periodic case is deemed the most elegant because of its symmetry, and it arguably best models the spatially infinite case in which boundaries are sufficiently far away to be of no concern.

Since a computer is a finite-state machine, it is necessary to represent the spatial variable x at finitely many discrete values, separated by a small distance k, much as we have already done with the time dependence for ODEs. Then we approximate the spatial derivatives at each grid point by fitting a smooth curve to the nearby points and taking its derivatives. To capture the nth derivative of u, we need a minimum of $n + 1$ points, so that, for example, the fourth derivative u_{xxxx} at $x = ik$ requires a minimum of five such points at $x_i, x_{i\pm1}$, and $x_{i\pm2}$, in which case the first four derivatives obtained by differentiating the fourth-order Lagrange interpolating polynomial for the five-point stencil (Abramowitz and Stegun, 1970) are given by

$$u_x = (u_{i-2} - 8u_{i-1} + 8u_{i+1} - u_{i+2})/12k$$

$$u_{xx} = (-u_{i-2} + 16u_{i-1} - 30u_i + 16u_{i+1} - u_{i+2})/12k^2$$

$$u_{xxx} = (-u_{i-2} + 2u_{i-1} - 2u_{i+1} + u_{i+2})/2k^3$$

$$u_{xxxx} = (u_{i-2} - 4u_{i-1} + 6u_i - 4u_{i+1}u + u_{i+2})/k^4.$$

(8.1)

With these approximations, the method can be considered as a fourth-order spatial discretization of the spatial dependence, much as the Runge–Kutta method that was used for solving the ODEs in the previous chapters is a fourth-order *temporal* discretization. However, in the spatial case, we have the advantage of knowing the spatial values on both sides of the point to be iterated, whereas in the temporal case, we are always projecting forward in time. We will not consider spatial derivatives higher than the fourth (u_{xxxx}) since they would begin to lose elegance and would require a larger spatial neighborhood and higher-order fitting functions, further straining the already large computational demands of PDE solvers.

A straightforward extension of the methods previously used to solve coupled ODEs resulting from the above approximations is the *method of lines* (Schiesser, 1991) in which the ODE corresponding to each grid point u_i and coupled to its four nearest neighbors through Eq. (8.1) is solved in time using a highly accurate fourth-order Runge–Kutta integrator with adaptive step size. The method is illustrated in Fig. 8.1 for the 201 ODEs that are used to solve the Kuramoto–Sivashinsky equation in the next section. The time evolution of successive spatial values of u_i is shown with each line displaced vertically downward from the previous by $\Delta u = 1$.

Unfortunately, there are issues of numerical instability (Press *et al.*, 2007) that raise questions about the reliability of this (or any other) PDE solution method (Schiesser and Griffiths, 2009). However, if the time step is sufficiently small compared with the spatial step and the temporal order is at least as high as the spatial order (fourth in this case), any instability usually grows slowly enough that valid solutions can be obtained for a time long enough to exhibit the solution and calculate its Lyapunov exponent. When a numerical instability grows to the point where it affects the dynamics, it is usually obvious since the solution will become unbounded, or the Lyapunov exponent will change abruptly, or there will be significant spatial structure at the shortest wavelengths. Of course, the instability could be real rather than numerical, and so it is possible that some of the cases included in this chapter are only transiently chaotic, but any such transients are of very long duration.

Partial differential equations are characterized by boundary conditions as well as the initial conditions that characterize ODEs. Many different boundary conditions are possible, but we will consider only examples in which the boundary conditions are periodic such that $u(L,t) = u(0,t)$, $u_x(L,t) = u_x(0,t)$, and so forth, where $L = Nk$ and N is the number of ODEs used to approximate the PDE. Thus the dynamics are assumed to occur on a one-dimensional ring analogous to the cases in the previous chapter. With L sufficiently large, such a ring is intended to approximate an infinitely large system with no ends. We will typically take $L = 100$

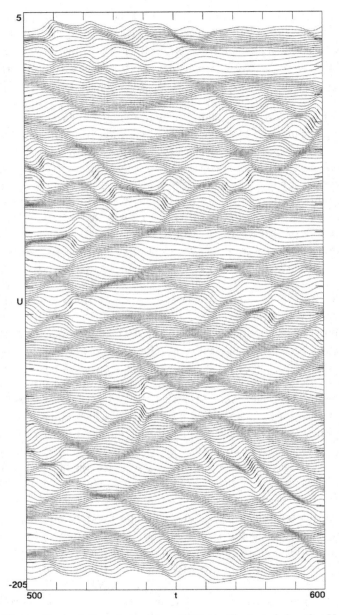

Fig. 8.1   The method of lines for solving PDEs is illustrated for the Kuramoto–Sivashinsky equation. The time evolution of successive spatial values of the 201 values of u_i is shown with each line displaced vertically downward from the previous by $\Delta u = 1$.

and $N = 201$ to be large but computationally tractable. This choice gives $k = L/N \approx 0.5$. The periodicity translates to the conditions $u_{N+1} = u_1$, $u_0 = u_N$, and so forth as in the previous chapter.

The methods and examples in this chapter are similar to those in the previous chapter except that the elegance is optimized for the PDE rather than for the ODEs that approximate it, the spatial derivatives are better approximated, more ODEs are used with a smaller spatial step size, and cases are avoided in which the chaos would obviously disappear in the continuum limit or where the PDE would otherwise exhibit pathological behavior.

There are other methods for solving PDEs, especially when the boundary conditions are periodic. For example, one could use a *spectral method* (Canuto *et al.*, 2006) in which the solution is represented as a sum of sinusoidal Fourier components, each containing an integer number of wavelengths, $u(x,t) = \sum_{i=1}^{N} a_i(t)\sin(2\pi ix/L) + b_i(t)\cos(2\pi ix/L)$, which converts the PDE into a coupled system of $2N$ ODEs with variables $a_i(t)$ and $b_i(t)$, much as does the previous method. Equivalently, one can represent the solution in the complex plane with N complex coefficients $c_i(t)$. The nonlinearities introduce spatial harmonics that are typically ignored once their wavelengths become shorter than $k = L/N$. The variables in the ODEs, $c_i(t)$ or $a_i(t)$ and $b_i(t)$, are amplitudes of the Fourier components, which must be recombined after finding their solution to obtain an approximation to $u(x,t)$.

8.2 Kuramoto–Sivashinsky Equation

One of the simplest PDEs that is known to exhibit chaos is the *Kuramoto–Sivashinsky equation* given by

$$u_t = -uu_x - u_{xx} - u_{xxxx}. \tag{8.2}$$

The u_{xx} term is a negative viscosity leading to the growth of long wavelength modes, and the u_{xxxx} term is a hyperviscosity that damps the short wavelength modes. The nonlinearity uu_x transports energy from the growing modes to the damped modes. This equation has been used to model waves in chemical reactions (Kuramoto and Tsuzuki, 1976), flame front modulations (Sivashinsky, 1977), and a thin liquid film flowing down an inclined plane (Sivashinsky and Michelson, 1980), and it has been extensively studied both analytically (Kudryashov, 1990) and numerically (Kevrekidis *et al.*, 1990). Its spatiotemporal plot as shown in Fig. 8.2 resembles that

Elegant Chaos

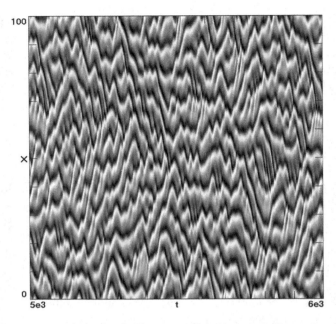

Fig. 8.2 Spatiotemporal plot for the Kuramoto–Sivashinsky equation system (case PD_8 in Table 8.1) in Eq. (8.2) with $L = 100$.

for the hyperviscous ring in Fig. 7.12, which contains the same terms as the discretized Kuramoto–Sivashinsky equation but in different magnitudes because it was optimized for elegance of the system of ODEs rather than for the PDE that it approximates. In this figure and the others in this chapter, the gray scale indicates the value of $u(x, t)$ at each point in space and time, with black representing the smallest value of u and white representing the largest value.

8.3 Kuramoto–Sivashinsky Variants

In an extensive search, Brummitt and Sprott (2009) were unable to find chaos in the 210 PDEs with one quadratic nonlinearity and the 163 PDEs with one cubic nonlinearity of the form $u_t = f(u, u_x, u_{xx}, \ldots)$ that are algebraically simpler than the Kuramoto–Sivashinsky equation. However, there are variants of the Kuramoto–Sivashinsky equation that are only slightly more complicated and that exhibit chaos.

8.3.1   *Cubic case*

One such example with a cubic nonlinearity is

$$u_t = -u^2 u_x - u_{xx} - u_{xxxx}. \tag{8.3}$$

Its spatiotemporal plot as shown in Fig. 8.3 shows an approximate spatial and temporal periodicity corresponding to a traveling wave with a phase velocity of order unity, but with a chaotic modulation.

8.3.2   *Quartic case*

An example of a Kuramoto–Sivashinsky variant with a quartic nonlinearity is

$$u_t = -u^3 u_x - u_{xx} - u_{xxxx}. \tag{8.4}$$

Its spatiotemporal plot as shown in Fig. 8.4 shows rapidly propagating, nearly periodic structures with a weak chaotic modulation.

8.4   Chaotic Traveling Waves

The previous Kuramoto–Sivashinsky variants resemble traveling waves with a weak chaotic modulation. Li (2004) has described a method whereby PDEs with traveling wave solutions can be converted into low-dimensional systems of ODEs, many of which resemble the jerk systems in previous chapters and are thus chaotic. The method works as follows: Assume the solution $u(x,t) = U(\xi)$ is a function of $\xi = x - ct$ (a wave traveling in the $+x$ direction with velocity c). Then the partial derivatives of $u(x,t)$ are given by $u_t = -cU'$, $u_x = U'$, $u_{xx} = U''$, and so forth, where the prime ($'$) denotes a derivative $U' = dU/d\xi$ with respect to ξ.

As an example, we can apply the method to the Kuramoto–Sivashinsky equation in Eq. (8.2) to obtain

$$U'''' + U'' + U'(U - c) = 0. \tag{8.5}$$

This fourth-order ODE (hyperjerk system) can be integrated once to get a third-order ODE

$$U''' + U' + U(U/2 - c) = 0, \tag{8.6}$$

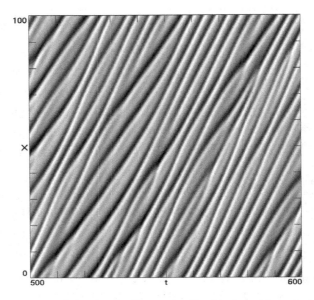

Fig. 8.3 Spatiotemporal plot for the Kuramoto–Sivashinsky cubic variant (case PD_{10} in Table 8.1) in Eq. (8.3) with $L = 100$.

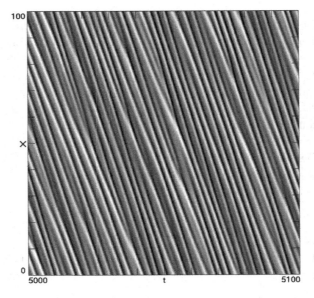

Fig. 8.4 Spatiotemporal plot for the Kuramoto–Sivashinsky quartic variant (case PD_{11} in Table 8.1) in Eq. (8.4) with $L = 100$.

which is a conservative jerk system but not one of those found to be chaotic in Table 4.1, even if one adds a constant of integration. This result implies that the Kuramoto–Sivashinsky equation does not have chaotic traveling wave solutions, and indeed Fig. 8.2 does not suggest propagating structures. The absence of chaos in the corresponding ODE thus does not imply the absence of chaos in the PDE from which it was derived since a particular and somewhat restrictive form of the solution was assumed. Notice that to have chaotic traveling wave solutions, the corresponding ODE must contain at least a third derivative U''', which implies at least a third derivative u_{xxx} in the PDE.

However, Fig. 8.3 does exhibit chaotic propagating structures, and so we would expect the corresponding ODE derived from Eq. (8.3) and given by

$$U''' + U' + U\left(U^2/3 - c\right) = 0 \tag{8.7}$$

to have chaotic solutions. Indeed, this equation is of the same form as Model CJ_1 in Table 4.1. Curiously, this ODE is conservative, whereas the PDE from which it was derived is dissipative.

8.4.1 *Rotating Kuramoto–Sivashinsky system*

It is possible to apply the method in reverse to generate chaotic PDEs from known systems of chaotic ODEs. Furthermore, it turns out that infinitely many PDEs can be constructed from a single ODE by the above method because U' can be represented either as $-u_t/c$ or as u_x or as a linear combination of the two. For example, the Kuramoto–Sivashinsky system in Eq. (8.2) can be generalized by adding a linear u_x term,

$$u_t = -uu_x - u_x - u_{xx} - u_{xxxx}, \tag{8.8}$$

which causes the system to rotate as shown in Fig. 8.5. Note that the Lyapunov exponent is essentially the same as for Fig. 8.2, suggesting that the added term merely produces rotation without otherwise altering the dynamics. The dynamics in the frame of reference rotating with velocity c are the same as the dynamics in the stationary reference frame.

8.4.2 *Rotating Kuramoto–Sivashinsky variant*

The previous system can be further generalized by adding a term linear in u_{xxx},

$$u_t = -uu_x - u_x - u_{xx} + 0.1u_{xxx} - u_{xxxx}, \tag{8.9}$$

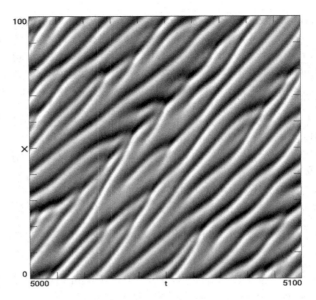

Fig. 8.5 Spatiotemporal plot for the rotating Kuramoto–Sivashinsky system (case PD_{12} in Table 8.1) in Eq. (8.8) with $L = 100$.

which suppresses but does not eliminate the chaos provided the coefficient of the u_{xxx} term (0.1) is not too large. The resulting spatiotemporal plot in Fig. 8.6 is a bit smoother and more regular than the fully chaotic one in Fig. 8.5.

8.5 Continuum Ring Systems

Another way to construct chaotic PDEs is to take the continuum limit of the chaotic ring systems in Chapter 7.

8.5.1 *Quadratic ring system*

One simple such case that involves only a first derivative in time, only the two nearest neighbors, and only quadratic nonlinearities is Eq. (7.5). The lowest two spatial derivatives at $x = ik$ can be represented by three-point approximations with $k = 0.5$, so that $u_x = u_{i+1} - u_{i-1}$ and $u_{xx} = 4u_{i+1} - 8u_i + 4u_{i-1}$. Substitution into Eq. (7.5) then leads directly to the conservative system

$$u_t = (1 - u_{xx} - 2u)u_x, \qquad (8.10)$$

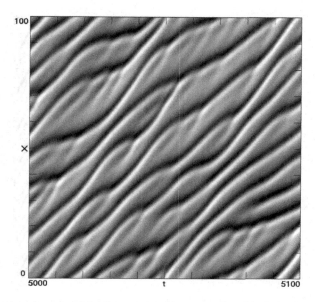

Fig. 8.6 Spatiotemporal plot for the rotating Kuramoto–Sivashinsky variant (case PD_{14} in Table 8.1) in Eq. (8.9) with $L = 100$.

which is then solved using the more accurate approximations to the derivatives in Eq. (8.1), resulting in the weakly chaotic spatiotemporal plot shown in Fig. 8.7.

8.5.2 *Antisymmetric quadratic system*

An even simpler example follows from Eq. (7.4), which aside from a scaling factor on the time, according to Eq. (8.1) is exactly equivalent to

$$u_t = uu_x. \tag{8.11}$$

This is a conservative system representing the zero-dispersion limit of the Korteweg–de Vries (KdV) equation (Korteweg and de Vries, 1895), and it has been investigated by Lax and Levermore (1983). Its spatiotemporal plot as shown in Fig. 8.8 not surprisingly resembles Fig. 7.6. The highly localized structures are chaotic *solitons* (nonlinear waves) that neither damp nor disperse as they propagate and that can pass through one another without distortion and with only a slight time lag. There is structure at the smallest resolved spatial scale, and indeed the filaments that thread through the plot become increasingly narrow as the number N of approxi-

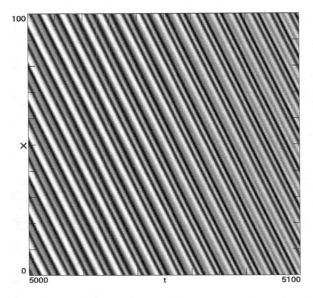

Fig. 8.7   Spatiotemporal plot for the quadratic continuum ring system (case PD_6 in Table 8.1) in Eq. (8.10) with $L = 100$.

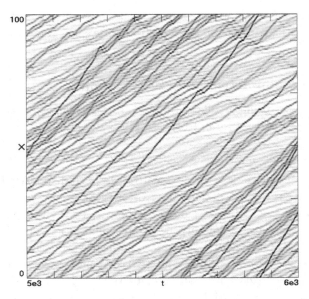

Fig. 8.8   Spatiotemporal plot for the antisymmetric quadratic continuum system (case PD_1 in Table 8.1) in Eq. (8.11) with $L = 100$.

mating ODEs increases, suggesting that they would be infinitely narrow in the true continuum limit.

Natural solitons were first observed in 1834 by John Scott Russell (Russell, 1834) while riding on horseback alongside the Union Canal near Edinburgh (Darrigol, 2003). They have been observed in the solution of a variety of PDEs, and there is an extensive literature devoted to them. See, for example, Drazin and Johnson (1989) and Ablowitz and Clarkson (1991). The term 'soliton' was coined by Zabusky and Kruskal (1965).

8.5.3 *Other simple PDEs*

An extensive but not exhaustive search for other chaotic systems of the form $u_t = f(u, u_x, u_{xx}, u_{xxx}, u_{xxxx})$, where f is a polynomial function of u and its spatial derivatives up to the fourth, turned up a number of interesting candidates as listed in Table 8.1. Some of the cases have been previously described. They are arranged in order of increasing complexity as defined by Brummitt and Sprott (2009), and they may not all survive close scrutiny in the limit as the number of approximating ODEs approaches infinity, but they are offered here as candidate systems for further study. Because of the large amount of computation required, these cases and others in this chapter are not necessarily optimally elegant.

The spatiotemporal plots for those cases not previously described are shown in Figs. 8.9 through 8.17.

Table 8.1   Chaotic partial differential equations.

Model	Equation	$u_0(x)$	Ly Ex
PD_1	$u_t = uu_x$	$-0.1 + 0.1\sin(2\pi x/L)$	0.0148
PD_2	$u_t = uu_{xxx}$	$0.1 - 0.1\sin(8\pi x/L)$	0.1610
PD_3	$u_t = (u-1)u_x$	$\sin(2\pi x/L)$	0.1488
PD_4	$u_t = (u+1)u_{xxx}$	$\sin(4\pi x/L)$	0.3870
PD_5	$u_t = (u+1)u_x - 0.1u_{xxx}$	$0.25\sin(4\pi x/L)$	0.0236
PD_6	$u_t = (1 - u_{xx} - 2u)u_x$	$-0.55 + 0.04\sin(14\pi x/L)$	0.0033
PD_7	$u_t = (0.05u_{xxxx} - u^2)u_x$	$2 + 0.2\sin(16\pi x/L)$	0.0240
PD_8	$u_t = -uu_x - u_{xx} - u_{xxxx}$	$\sin(2\pi x/L)$	0.0902
PD_9	$u_t = (u_{xxx} - u_{xx}^2)u$	$0.1\sin(16\pi x/L)$	0.0017
PD_{10}	$u_t = -u^2u_x - u_{xx} - u_{xxxx}$	$1 + \sin(18\pi x/L)$	0.0678
PD_{11}	$u_t = -u^3u_x - u_{xx} - u_{xxxx}$	$\sin(2\pi x/L)$	0.0852
PD_{12}	$u_t = -uu_x - u_x - u_{xx} - u_{xxxx}$	$\sin(2\pi x/L)$	0.0905
PD_{13}	$u_t = 1 + (1 - u^3u_{xxx})u_{xxx}$	$\sin(16\pi x/L)$	4.3077
PD_{14}	$u_t = -(u+1)u_x - u_{xx} + 0.1u_{xxx} - u_{xxxx}$	$\sin(2\pi x/L)$	0.0604
PD_{15}	$u_t = -(u_{xxx}u_{xxxx} + u)u_{xxx}$	$\sin(24\pi x/L)$	0.0105
PD_{16}	$u_t = (u_x^2 + u_{xx}u_{xxx})u_{xxx} - 0.1u_{xx}$	$\sin(2\pi x/L)$	0.4902

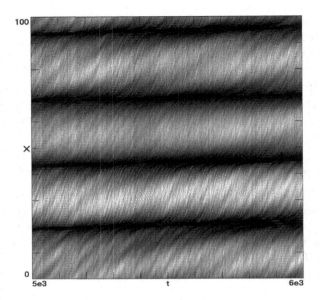

Fig. 8.9   Spatiotemporal plot for the chaotic PDE system PD_2 in Table 8.1 with $L = 100$.

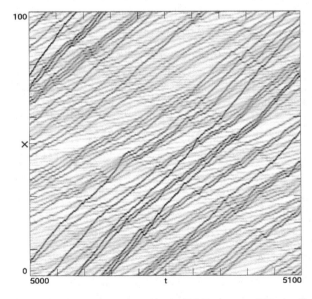

Fig. 8.10   Spatiotemporal plot for the chaotic PDE system PD_3 in Table 8.1 with $L = 100$.

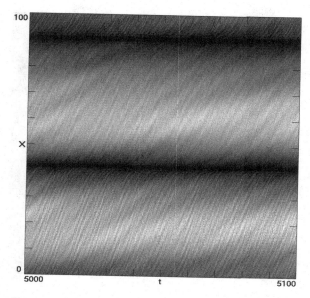

Fig. 8.11 Spatiotemporal plot for the chaotic PDE system PD_4 in Table 8.1 with $L = 100$.

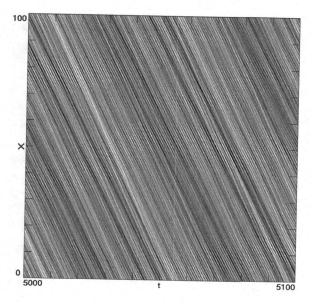

Fig. 8.12 Spatiotemporal plot for the chaotic PDE system PD_5 in Table 8.1 with $L = 100$.

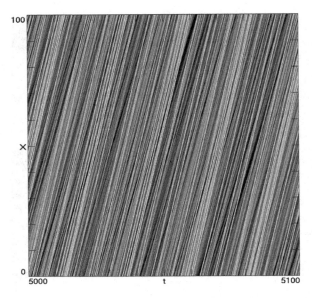

Fig. 8.13   Spatiotemporal plot for the chaotic PDE system PD_7 in Table 8.1 with $L = 100$.

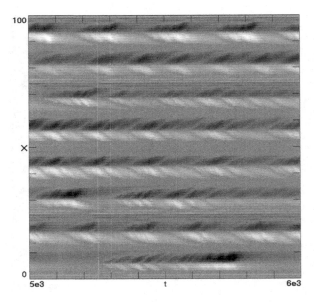

Fig. 8.14   Spatiotemporal plot for the chaotic PDE system PD_9 in Table 8.1 with $L = 100$.

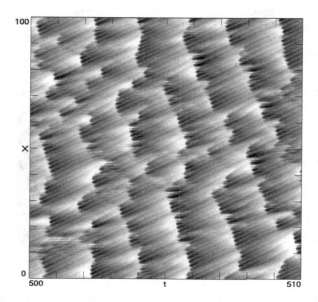

Fig. 8.15   Spatiotemporal plot for the chaotic PDE system PD_{13} in Table 8.1 with $L = 100$.

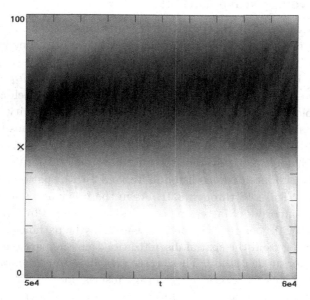

Fig. 8.16   Spatiotemporal plot for the chaotic PDE system PD_{15} in Table 8.1 with $L = 100$.

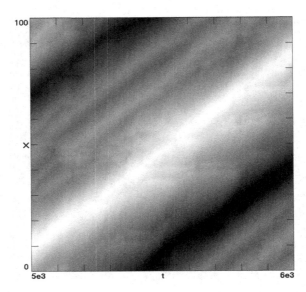

Fig. 8.17   Spatiotemporal plot for the chaotic PDE system PD_{16} in Table 8.1 with $L = 100$.

8.6   Traveling Wave Variants

A logical extension of the PDEs in the previous sections is to consider systems of that involve the second time derivative of u, one simple form of which is $u_{tt} = f(u, u_x, u_{xx}, u_{xxx}, u_{xxxx}, u_t)$, where f is a polynomial function of its arguments. The PDE for a traveling wave $u_{tt} = u_{xx}/c^2$ is of this form, having a solution $u(x, t) = A \sin(kx - \omega t)$ which corresponds to a sinusoidal wave of amplitude A traveling with a velocity $c = \omega/k$ in the $+x$ direction. Of course such a wave cannot exhibit chaos because the governing PDE is linear.

A search for chaotic systems of the above general form turned up a number of candidate cases as listed in Table 8.2 with spatiotemporal plots as shown in Figs. 8.18 through 8.31. As in the previous table, these cases are arranged in order of increasing complexity as defined by Brummitt and Sprott (2009). As before, some of these cases may be pathological, with the chaos resulting from the spatial discretization, but they are offered here as candidates for further study. These cases are called *traveling wave variants* because they contain a term linear in u_{tt} just as does the linear traveling wave equation, which also has a term linear in u_{xx}. As with the previous cases, these examples are not necessarily optimally elegant.

Table 8.2   Chaotic traveling wave PDE variants.

Model	Equation	$u_0(x)$ with $\dot{u}_0(x) = 0$	Lyap Exp
TW_1	$u_{tt} = -u^3 + u_{xx}$	$2 - \sin(24\pi x/L)$	0.0895
TW_2	$u_{tt} = -u^2 - u_{xxxx}$	$1 + 0.2\sin(6\pi x/L)$	0.0353
TW_3	$u_{tt} = -(u^2 + 1)u - u_{xx}$	$\sin(4\pi x/L)$	1.0916
TW_4	$u_{tt} = -(u^2 + 1)u - u_{xxxx}$	$1 - \sin(18\pi x/L)$	0.0193
TW_5	$u_{tt} = -u^3 + 0.1uu_t - u_{xx}$	$\sin(4\pi x/L)$	0.7703
TW_6	$u_{tt} = u^2 u_{xx} + u_{xxx} - u_t$	$\sin(8\pi x/L)$	1.2933
TW_7	$u_{tt} = u_{xx}^2 - 0.5u - u_{xxxx}$	$\sin(20\pi x/L)$	0.0020
TW_8	$u_{tt} = (u_{xx}^2 - 1)u_{xx} - u$	$\sin(2\pi x/L)$	1.1107
TW_9	$u_{tt} = uu_{xx}u_{xxxx} - 0.4u_{xxx} - u_t$	$\sin(2\pi x/L)$	0.6560
TW_{10}	$u_{tt} = -u_{xxxx}^3 + 1 - u$	$1 + \sin(34\pi x/L)$	1.1172
TW_{11}	$u_{tt} = u_x^2 u_{xx} - u_{xxx} - u_{xxxx} - u_t$	$\sin(8\pi x/L)$	0.7982
TW_{12}	$u_{tt} = u_{xx}u_{xxxx}^2 - u_{xxx} - u_t$	$\sin(18\pi x/L)$	0.5375
TW_{13}	$u_{tt} = (u_{xxxx}^2 - 2)u_{xxx} - 2u_{xxxx} - u_t$	$\sin(20\pi x/L)$	0.1002

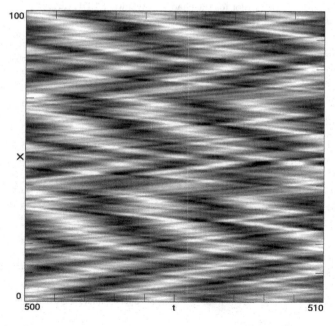

Fig. 8.18   Spatiotemporal plot for the chaotic travel wave PDE variant TW_1 in Table 8.2 with $L = 100$.

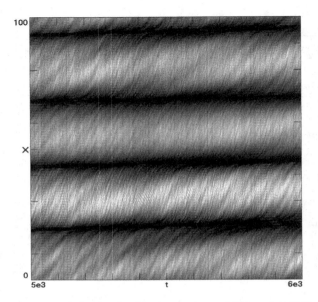

Fig. 8.19   Spatiotemporal plot for the chaotic travel wave PDE variant TW_2 in Table 8.2 with $L = 100$.

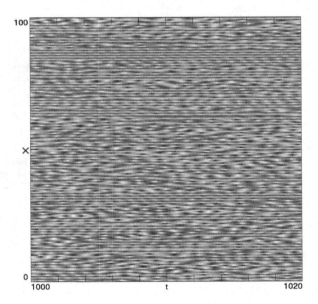

Fig. 8.20   Spatiotemporal plot for the chaotic travel wave PDE variant TW_3 in Table 8.2 with $L = 100$.

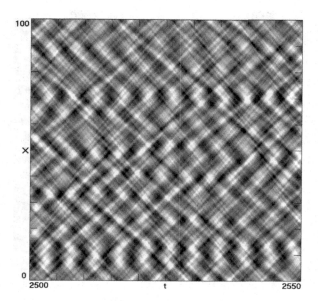

Fig. 8.21 Spatiotemporal plot for the chaotic travel wave PDE variant TW_4 in Table 8.2 with $L = 100$.

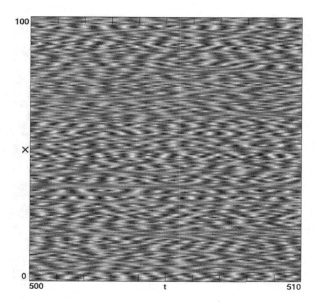

Fig. 8.22 Spatiotemporal plot for the chaotic travel wave PDE variant TW_5 in Table 8.2 with $L = 100$.

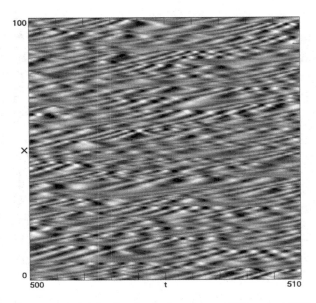

Fig. 8.23   Spatiotemporal plot for the chaotic travel wave PDE variant TW_6 in Table 8.2 with $L = 100$.

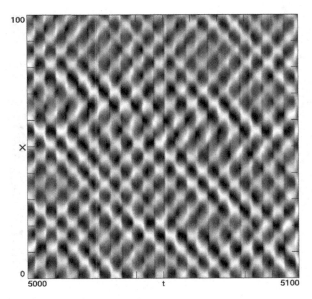

Fig. 8.24   Spatiotemporal plot for the chaotic travel wave PDE variant TW_7 in Table 8.2 with $L = 100$.

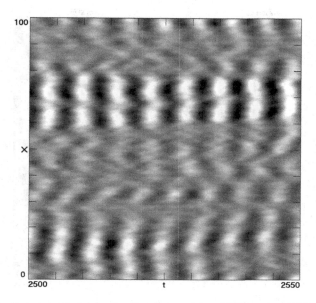

Fig. 8.25 Spatiotemporal plot for the chaotic travel wave PDE variant TW_8 in Table 8.2 with $L = 100$.

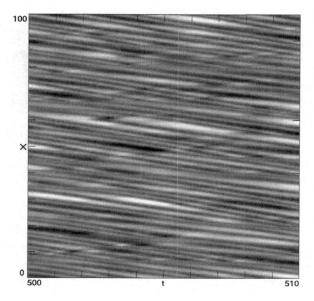

Fig. 8.26 Spatiotemporal plot for the chaotic travel wave PDE variant TW_9 in Table 8.2 with $L = 100$.

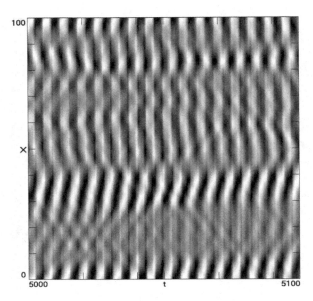

Fig. 8.27 Spatiotemporal plot for the chaotic travel wave PDE variant TW_{10} in Table 8.2 with $L = 100$.

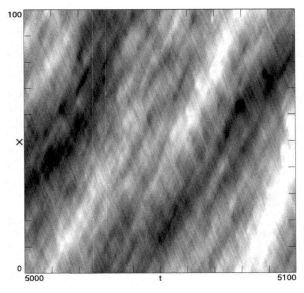

Fig. 8.28 Spatiotemporal plot for the chaotic travel wave PDE variant TW_{11} in Table 8.2 with $L = 100$.

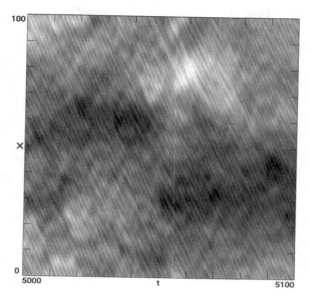

Fig. 8.29 Spatiotemporal plot for the chaotic travel wave PDE variant TW_{12} in Table 8.2 with $L = 100$.

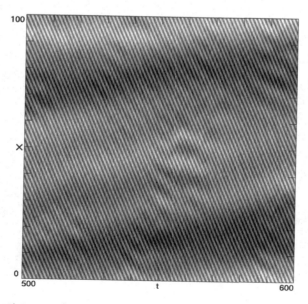

Fig. 8.30 Spatiotemporal plot for the chaotic travel wave PDE variant TW_{13} in Table 8.2 with $L = 100$.

Chapter 9

Time-Delay Systems

Another class of system that is infinite-dimensional and that can exhibit chaos with a relatively simple model involves the value of the dynamical variable at one or more times in the past. Such systems are relatively unexplored but are reasonable models for many important processes, especially in biology. They provide elegant examples of chaos in systems that are much too simple to exhibit chaos without the time delay.

9.1   Delay Differential Equations

There are many ways to introduce time delays into a dynamical system, but we will here consider simple examples of *delay differential equations* (DDEs) (Kuang, 1993; Erneux, 2009), also called *differential delay equations* (DDEs), *retarded delay differential equations* (RDDEs), or *retarded functional differential equations* (RFDEs)(Hale, 1977), which are a special type of *functional differential equation* (FDE) in which the past dependence is through the single real state space variable rather than through its derivatives. In addition, the dependence will be autonomous (not explicitly involving time) and will involve the value of the state variable at a single discrete time lag as well as the current time as given by

$$\dot{x} = f(x, x_\tau). \tag{9.1}$$

The notation x_τ will be used to denote the value of x at an earlier time of $t - \tau$. This is a *first-order* DDE since it contains only the first time derivative of x, and it is *explicit* since the $\dot{x}$ term is by itself on the left of the equal sign.

It is important to recognize that there is a multitude of other DDEs including ones that involve multiple time lags or a weighted integral over all previous lags, ones in which the delay time is itself a function of the variables, ones in which the time derivative appears implicitly in the equation, ones in which time derivatives higher than the first are present, ones involving fractional derivatives, ones with multiple variables, ones in which the variables are complex, and ones that involve an explicit time dependence (nonautonomous). Furthermore, partial differential equations can include time delays (DPDEs). However, in the spirit of this book, only the simplest case will be considered, except for one example of a DDE with a continuous delay in the final section, but the nonlinearity will have a variety of elementary forms.

DDEs are most easily solved using the *forward Euler method* (Sprott, 2003), especially when f depends only on x_τ since the Euler method entails an inevitable time lag on the order of $h/2$. The cases presented in this chapter were solved by that method with $N = 1 \times 10^4$ and $h = \tau/(N - 1/2)$. DDEs require a continuum of initial conditions specifying the values of x for all times from $t = -\tau$ to $t = 0$, which translates into N initial conditions for the discrete-time approximation. We will typically take these initial conditions to be a constant x_0. Other methods for solving DDEs are described by Cryer (1972) and by Bellen and Zennaro (2003), and an extension of the Runge–Kutta method to DDEs is given by Virk (1985).

An interesting feature of DDEs is that the initial condition does not in general satisfy the equation. In particular, the first derivative will usually have a discontinuity at $t = 0$ since the value of $\dot{x}$ just after $t = 0$ will be determined by the equation, whereas the value of $\dot{x}$ just before $t = 0$ will be determined by the initial conditions. However, the equation will usually smooth out any discontinuities in the initial condition as time goes on, especially when the system is dissipative. In particular, the discontinuity at $t = 0$ will propagate forward until $t = \tau$, whereupon the first derivative $\dot{x}$ becomes continuous, but the second derivative $\ddot{x}$ will be discontinuous, and so forth until only derivatives higher than t/τ are problematic. Thus the same strange attractor is obtained for a wide range of initial conditions in a chaotic DDE once the initial transient has decayed. The same smoothing of the initial conditions also occurs in conservative DDEs despite the absence of an attractor.

DDEs have been used extensively to model population dynamics (Kuang, 1993) with their inherent maturation and gestation time delays, but also to study epidemics (Sharpe and Lotka, 1923), tumor growth (Villasana and Radunskaya, 2003), immune systems (Nelson and Perelson, 2002), lossless electrical transmission lines (Brayton, 1966), and the electrodynamics of charged particles interacting through the Lorenz force with Linard–Weichert potentials (Driver, 1984), among many others (Kolmanovski and Myshkis, 1999).

9.2   Mackey–Glass Equation

One of the earliest and most widely studied DDE is the *Mackey–Glass equation* (Mackey and Glass, 1977; Farmer, 1982; Grassberger and Procaccia, 1984),

$$\dot{x} = \frac{ax_\tau}{1 + x_\tau^c} - bx, \tag{9.2}$$

proposed to model the production of white blood cells, but it could also apply to the spread of a disease with an incubation period or the growth of an animal population with a gestation period. Mackey and Glass (1977) showed that this equation exhibits chaos for $a = 0.2$, $b = 0.1$, $c = 10$, and $\tau = 23$ with an attractor as shown in Fig. 9.1. In this and the other figures in this chapter, the value of x at each time is plotted versus its value x_τ for a time delay τ with the third dimension represented by a shadow indicating the value of x at a time of $\tau/2$.

A slightly more elegant set of parameters for the Mackey–Glass equation giving a much larger Lyapunov exponent is $a = 3$, $b = 1$, $c = 7$, and $\tau = 3$, which gives the attractor shown in Fig. 9.2 that is rather similar to the one in Fig. 9.1.

9.3   Ikeda DDE

Another example of a chaotic DDE was proposed by Ikeda and Matsumoto (1987) to model a passive optical bistable resonator system (Ikeda, 1979) and is given by

$$\dot{x} = \mu \sin(x_\tau - c) - x. \tag{9.3}$$

With $\mu = 6$, $c = 0$, and $\tau = 1$, it gives the attractor shown in Fig. 9.3.

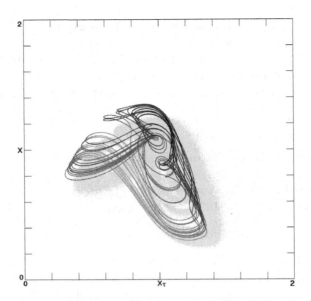

Fig. 9.1 Attractor for the Mackey–Glass equation in Eq. (9.2) with $(a, b, c, \tau) = (0.2, 0.1, 10, 23)$ and $x_0 = 0.9$, $\lambda = 0.0093$.

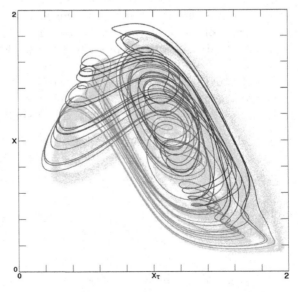

Fig. 9.2 Attractor for the Mackey–Glass equation in Eq. (9.2) with $(a, b, c, \tau) = (3, 1, 7, 3)$ and $x_0 = 1$, $\lambda = 0.0674$.

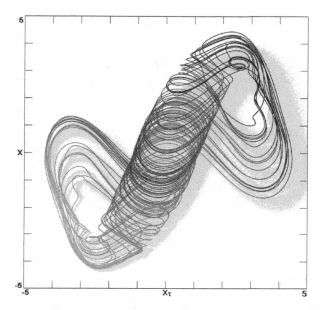

Fig. 9.3 Attractor for the Ikeda DDE in Eq. (9.3) with $(\mu, c, \tau) = (6, 0, 1)$ and $x_0 = 0.1$, $\lambda = 0.4285$.

9.4 Sinusoidal DDE

It turns out that the Ikeda DDE can be simplified even further (Sprott, 2007b) giving the almost trivially simple system

$$\dot{x} = \sin x_\tau \qquad (9.4)$$

whose only parameter is τ. It exhibits chaos for most values of $\tau > 4.991$ except for a (possibly infinite) number of periodic windows. Its state space plot for the weakly chaotic case with $\tau = 5$ is shown in Fig. 9.4.

9.5 Polynomial DDE

An expansion of Eq. (9.4) about $x_\tau = 0$ (where $\sin x_\tau \approx x_\tau - x_\tau^3/6$) leads one to suspect that a polynomial such as

$$\dot{x} = x_\tau - x_\tau^3 \qquad (9.5)$$

might exhibit chaos. Indeed, it does for most values of τ in the relatively narrow range of $1.538 < \tau < 1.723$. Figure 9.5 shows its state space plot

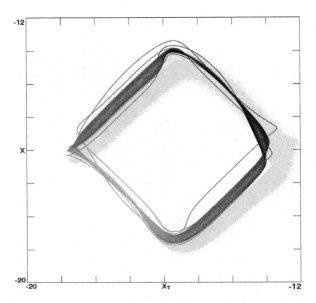

Fig. 9.4   State space plot for the sinusoidal DDE in Eq. (9.4) with $\tau = 5$ and $x_0 = 0.1$, $\lambda = 0.0330$.

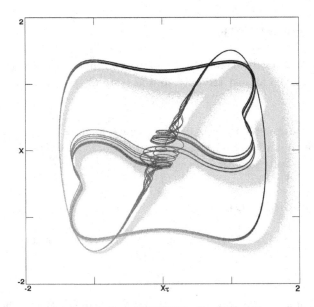

Fig. 9.5   State space plot for the sinusoidal DDE in Eq. (9.5) with $\tau = 1.7$ and $x_0 = 0.1$, $\lambda = 0.0818$.

for $\tau = 1.7$. Similar behavior is observed with the signs interchanged in Eq. (9.5) for values of τ reaching $\tau = 3.815$ before the solution becomes unbounded.

Other simple polynomial DDEs include the *logistic delay differential equation* (Losson et al., 1993) $\dot{x} = \lambda x_\tau (1 - x_\tau) - \alpha x$ and *Hutchinson's equation* $\dot{x} = \alpha x (1 - x_\tau)$, both used to model single-species population growth (May, 1975), *Wright's equation* (Wright, 1955) $\dot{x} = \alpha x_\tau (1 - x)$ used in number theory for predicting the distribution of prime numbers, and the *delayed-action oscillator* (Suarez and Schopf, 1988) $\dot{x} = x(1 - x^2) - \alpha x_\tau$ used to model El Niño oscillations in the equatorial Pacific, each of which admits periodic oscillations but apparently not chaos. Thus Eq. (9.5) may be the simplest chaotic polynomial DDE.

9.6 Sigmoidal DDE

A related form involves a sigmoidal nonlinearity such as the hyperbolic tangent, an example of which that admits chaos is

$$\dot{x} = 2 \tanh x_\tau - x_\tau. \tag{9.6}$$

This case could be considered as an artificial neural network with a single neuron and a time delay (Marcus and Westervelt, 1989). Its state space plot for $\tau = 3$ is shown in Fig. 9.6.

9.7 Signum DDE

As is often the case, the hyperbolic tangent can be replaced with the signum nonlinearity, which has a similar behavior when the argument is large but that switches abruptly as the argument crosses zero. The simplest such form that admits chaos is (Yalçin and Özoguz, 2007)

$$\dot{x} = \operatorname{sgn} x_\tau - x_\tau. \tag{9.7}$$

As usual, for numerical reasons, the $\operatorname{sgn} x$ has been approximated by $\tanh(500x)$ for estimating the Lyapunov exponent. The state space plot for this case with $\tau = 2$ is shown in Fig. 9.7.

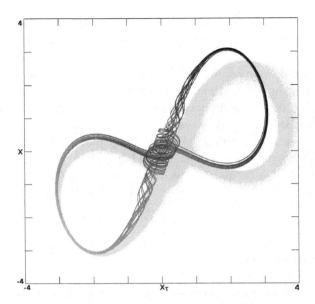

Fig. 9.6   State space plot for the sigmoidal DDE in Eq. (9.6) with $\tau = 3$ and $x_0 = 1$, $\lambda = 0.0618$.

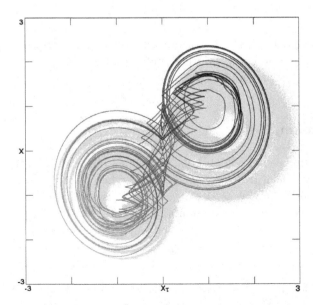

Fig. 9.7   State space plot for the signum DDE in Eq. (9.7) with $\tau = 2$ and $x_0 = 0.1$, $\lambda = 0.0627$.

9.8 Piecewise-linear DDEs

9.8.1 *Antisymmetric case*

A characteristic of most of the previous systems is that $df(x_\tau)/dx_\tau$ is positive for small $|x_\tau|$ and negative for large $|x_\tau|$. This leads one to suspect that a piecewise-linear system such as

$$\dot{x} = |x_\tau + 1| - |x_\tau - 1| - x_\tau \qquad (9.8)$$

might be chaotic. Indeed it is for $\tau = 3$ as Fig. 9.8 shows. Such piecewise-linear systems are especially amenable to implementation with electronic circuits (an der Heiden and Mackey, 1982; Lu and He, 1996; Lu *et al*, 1998; Tamaševičius *et al.*, 2006) as the next chapter will show.

9.8.2 *Asymmetric case*

Most of the previous equations are antisymmetric about $x_\tau = 0$ ($f(x_\tau) = -f(-x_\tau)$) and thus have at least two bends in their $f(x_\tau)$ curve. One might

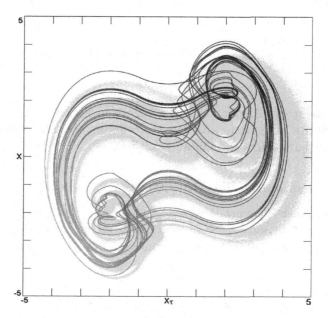

Fig. 9.8 State space plot for the piecewise-linear DDE in Eq. (9.8) with $\tau = 3$ and $x_0 = 0.1$, $\lambda = 0.0909$.

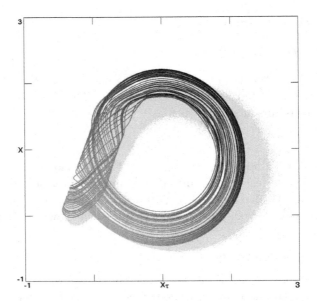

Fig. 9.9   State space plot for the piecewise-linear DDE in Eq. (9.9) with $\tau = 1.8$ and $x_0 = 0$, $\lambda = 0.0899$.

wonder if there are asymmetric piecewise-linear systems with a single bend in the curve. Indeed there are, one simple example of which is

$$\dot{x} = x_\tau - 2|x_\tau| + 1 \tag{9.9}$$

whose state space plot for $\tau = 1.8$ is shown in Fig. 9.9. This system is rather delicate in the sense that the chaos occurs over a relatively narrow range of its parameters and initial conditions.

9.8.3   *Asymmetric logistic DDE*

An even simpler asymmetric piecewise-linear case, motivated by its similarity to Eq. (9.5) and similar to the *logistic ordinary differential equation,* also called the *Verhulst equation* (Verhulst, 1838), is given by

$$\dot{x} = x_\tau(1 - |x_\tau|) \tag{9.10}$$

whose state space plot for $\tau = 3$ is shown in Fig. 9.10. Note that if we replace $|x_\tau||$ by x_τ^2 in Eq. (9.10), which has the same symmetry, we get exactly Eq. (9.5) which was previously shown to be chaotic.

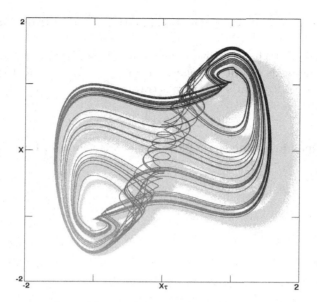

Fig. 9.10 State space plot for the asymmetric piecewise-linear logistic DDE in Eq. (9.10) with $\tau = 3$ and $x_0 = 0.9$, $\lambda = 0.0010$.

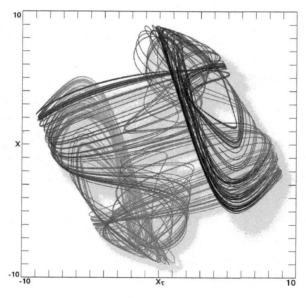

Fig. 9.11 State space plot for the asymmetric piecewise-linear logistic DDE with continuous delay in Eq. (9.11) with $\tau = 3$ and $x_0 = 1$, $\lambda = 0.1555$.

9.9   Asymmetric Logistic DDE with Continuous Delay

We conclude this chapter with one example of a DDE in which the time derivative depends on the average value of a function similar to Eq. (9.10) for time lags of x_s from $s = 0$ to τ according to

$$\dot{x} = \frac{1}{\tau} \int_0^\tau x_s(4 - |x_s|)ds. \tag{9.11}$$

The corresponding state space plot for $\tau = 3$ is shown in Fig. 9.11.

Chapter 10

Chaotic Electrical Circuits

There has been considerable interest in the design and construction of chaotic electrical circuits for their application to fields such as secure communications as well as to confirm that the computed solutions are not numerical artifacts. Such circuits can be elegant in their own right if they contain a small number of standard electronic components and are easy to construct. This concluding chapter describes a number of such circuits along with the equations that describe them and the predict ed electrical waveforms.

10.1 Circuit Elegance

Just as an equation is deemed elegant if it contains the fewest terms necessary to produce chaos and if the coefficients of those terms have simple integer values, a circuit can be considered elegant if it has the smallest number of standard electronic components and the values of those components are simple round numbers so that they can be easily and inexpensively procured. Chaos has been observed in very simple circuits such as the nonideal operational amplifier with a single feedback resistor described by Yim *et al.* (2004), but they rely upon hidden parasitic properties of the electronic components and are thus hard to model and reproduce. Such parasitic circuits are not very elegant and thus will not be discussed here.

Just as it is possible to transform a differential equation by linearly rescaling the time and values of the variables, it is usually possible to rescale a circuit by increasing or decreasing the values of the components to make it operate in the desired frequency range or to rescale the voltages and currents to exploit the properties of whatever nonlinear components are involved. Hence we will assume certain ideal components and let the reader

233

scale their values as required for the available components and the desired application.

Special integrated circuits are available to perform most of the mathematical functions described in the previous chapters. In particular, multiplier chips are available that accurately perform the operations required for any of the polynomial nonlinearities that have dominated the previous examples. Thus one could in principle construct a circuit to replicate the behavior of any of those systems. The circuit then just becomes an analog computer that solves the corresponding differential equations.

In this chapter we will rather be concerned with simple circuits that do not require such specialized components. We will restrict the discussion to circuits that contain the standard linear components — dc and sinusoidal voltage sources, resistors, capacitors, inductors, and linear operational amplifiers, as well as delay lines. However, chaos can occur only if the circuit contains a nonlinearity, and so we must also allow components that are not ideal such as saturating inductors and operational amplifiers as well as various kinds of diodes and their older gaseous counterparts. Optimizing the elegance then amounts to minimizing the component count.

Each class of system in the previous chapters has an electronic counterpart. In particular, we can distinguish sinusoidally forced circuits corresponding to the nonautonomous systems in Chapter 2 from the active circuits corresponding to the autonomous systems in the subsequent chapters. The former case will require an external oscillatory voltage source, while the latter will require one or more operational amplifiers with their associated dc power supply to overcome the inevitable resistive losses that accompany any real circuit. Just as we can convert a nonautonomous system of equations into an autonomous one by adding one or more variables, we can convert a periodically forced chaotic circuit into a self-contained chaotic oscillator either by adding a sinusoidal oscillator or an active element with positive feedback.

10.2   Forced Relaxation Oscillator

One of the earliest circuit in which chaotic oscillations were observed is the neon glow lamp forced relaxation oscillator studied by van der Pol and van der Mark (1927). Since little was known about chaos at that time, their interest was in *frequency demultiplication* in which the output of the circuit contains frequency components that are submultiples of the frequency Ω of

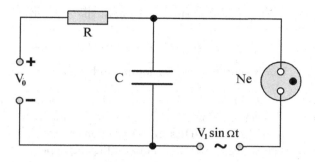

Fig. 10.1   Forced relaxation oscillator.

the sinusoidal input signal. This behavior is in contrast to the more familiar *frequency multiplication* that occurs through the generation of harmonics arising from nonlinearities. Using the circuit shown in Fig. 10.1, they were able to observe frequencies as low as $\Omega/200$ using 'a telephone coupled loosely in some way to the system.'

Unexpectedly, they also observed what we now call 'chaos' and wrote:

> Often an irregular noise is heard in the telephone receivers before the frequency jumps to the next lower value. However, this is a subsidiary phenomenon, the main effect being the regular frequency demultiplication.

Had they pursued that innocent observation, they would have discovered chaos half a century before it was otherwise widely known, and history might have been different.

The operation of the circuit is straightforward. Assume that the voltage V across the capacitor C is initially zero. The capacitor begins charging by the current $I = (V_0 - V)/R$ flowing through the resistor R. When the capacitor voltage reaches a sufficiently large value, the neon lamp abruptly begins to conduct and rapidly discharges the capacitor. The onset of current conduction depends on the value of the sinusoidal voltage $V_1 \sin \Omega t$ that is in series with the neon lamp. The name 'relaxation oscillator' comes from the fact that the capacitor charges slowly and then relaxes quickly to a lower voltage when the neon lamp begins to conduct. When the current through the neon lamp reaches a sufficiently low value, the conduction ceases, and the process begins again. The capacitor voltage $V(t)$ as a function of time thus resembles a sawtooth. In a modern implementation of the circuit, the neon lamp would be replaced by a diac or bidirectional breakover diode.

The equations that describe the variation of the capacitor voltage are

$$\dot{V} = (V_0 - V)/RC \quad \text{for lamp not conducting}$$
$$V = V_1 \sin \Omega t \qquad \text{for lamp conducting.}$$

$$(10.1)$$

Normally, a first order ODE would not be capable of producing chaos, even with a sinusoidal forcing. However, this circuit has memory (*hysteresis*), which makes it a bit like the DDEs in the previous chapter. In particular, the neon lamp is either conducting or not conducting depending on its past history.

We will assume an idealized nonlinear element that begins conducting when the voltage across its terminals $|V - V_1 \sin \Omega t|$ reaches exactly 1 Volt, then instantly brings the capacitor voltage to the voltage of the sinusoidal source, and immediately stops conducting. For a real device such as a neon lamp or diac with a different conduction threshold, one would scale V_0 and V_1 accordingly. A real neon lamp might typically begin conducting at 80 volts and stop conducting when the voltage falls below about 60 volts. The neon lamp would be observed to flash briefly each time the capacitor discharges, so that the chaos can be observed visually as irregular flashing without additional electronic instrumentation such as an oscilloscope.

Note also that R and C appear only as a product in Eq. (10.1), and that product has units of time. Thus without loss of generality, we can take $R = 1$ and $C = 1$, in which case the natural time scale is of order unity. Thus if $R = 1$ Ohm and $C = 1$ Farad, the time would be in seconds, although a more reasonable choice might be $R = 10^5$ Ohms and $C = 10^{-8}$ Farads so that the circuit currents would be more modest (milliamperes instead of Amperes) and the relaxation oscillation would be in the audible range (~ 1 kHz rather than ~ 1 Hz). Alternately, one could take $\Omega = 1$ and let RC be adjustable, which amounts to measuring time in units of $1/2\pi$ times the period of the forcing function. A typical chaotic waveform for the capacitor voltage in this circuit is shown in Fig. 10.2.

In this and the successive examples, the initial capacitor voltage (at $t = 0$) is assumed to be zero as would normally be the case when the circuit is first energized. The circuit is allowed to operate for a while (until $t = 100$ in this case, or roughly a hundred cycles) before plotting the output to ensure that the chaos is not just a transient. A value is not given for the Lyapunov exponent since the discontinuity in the voltage and the resulting theoretically infinite current when the neon lamp first conducts in this simple model makes it difficult to calculate a meaningful value.

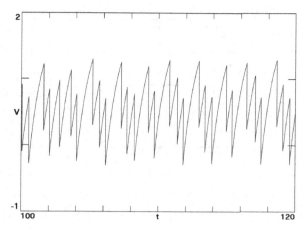

Fig. 10.2 Waveform for the capacitor voltage in the forced relaxation oscillator in Fig. 10.1 with $V_0 = 2$, $V_1 = 0.3$, $RC = 1$, and $\Omega = 3.3$.

10.3 Autonomous Relaxation Oscillator

Bernhardt (1991) pointed out that the sinusoidal voltage source in Fig. 10.1 can be replaced with a passive LC circuit as shown in Fig. 10.3. The operation is similar to the forced relaxation oscillator with the forcing frequency being determined by $\Omega = 1/\sqrt{LC_1}$. In the idealized circuit in Fig. 10.3, the LC circuit is assumed to have no dissipation (infinite Q), but in practice the Q can be at least as low as 33 (Bernhardt, 1991). Energy is supplied to the LC circuit each time the neon lamp fires, making up for any resistive losses, primarily in the inductor.

The equations that describe the circuit are

$$\dot{V} = (V_0 - V)/RC \quad \text{for lamp not conducting}$$

$$\dot{V}_1 = -I_L/C_1 \qquad \text{for lamp not conducting}$$

$$V = V_1 \qquad \qquad \text{for lamp conducting} \qquad \qquad (10.2)$$

$$\dot{I}_L = V_1/L$$

where V is the voltage across capacitor C and V_1 is the voltage across the capacitor V_1. The neon lamp is assumed to fire when $|V - V_1| > 1$, whereupon the voltage V instantly drops by an amount $\Delta V = -C_1/(C + C_1)$ and the voltage V_1 rises by an amount $\Delta V_1 = C/(C+C_1)$, and then the neon lamp ceases conducting. This brings the two capacitors momentarily

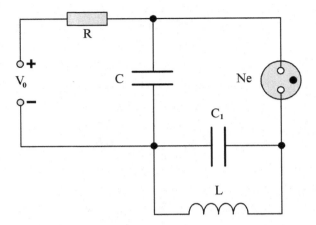

Fig. 10.3   Autonomous relaxation oscillator.

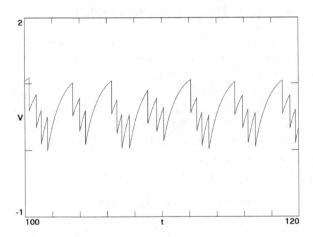

Fig. 10.4   Waveform for the voltage across the capacitor C in the autonomous relaxation oscillator in Fig. 10.3 with $V_0 = 1.2$, $R = C = C_1 = 1$, and $L = 0.09$.

to the same voltage. A typical waveform for $V(t)$ across the capacitor C is shown in Fig. 10.4.

This system is the electrical analog of the dripping faucet, which is known to be chaotic (Shaw, 1984). In the dripping faucet, the release of a water droplet corresponds to the firing of the neon lamp, and it causes the

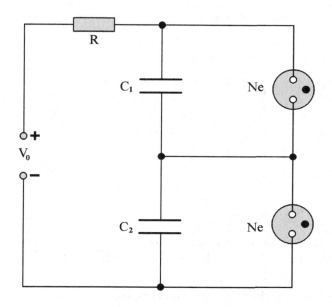

Fig. 10.5   Two coupled relaxation oscillators.

water left behind to oscillate, thereby affecting when the next drip occurs, much as the firing of the neon lamp excites oscillations in LC circuit, which affect when the lamp next fires.

10.4   Coupled Relaxation Oscillators

10.4.1   *Two oscillators*

Relaxation oscillators can also be coupled to produce chaos. One such circuit was described by Rolf Landauer in an internal IBM memorandum entitled 'Poor Man's Chaos' in 1977. A simplified version of his circuit (untested) is shown in Fig. 10.5. His circuit had two additional resistors, one in parallel with each capacitor, but in theory, those are not essential.

As with the previous circuits, in this idealization, the capacitors begin to charge through the resistor R until the voltage across one of the capacitors exceeds 1, and then the corresponding neon lamp fires, drops the voltage across its capacitor abruptly to zero, and then stops conducting.

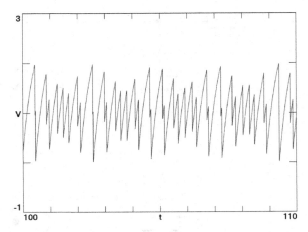

Fig. 10.6   Waveform for the total capacitor voltage $V(t) = V_1 + V_2$ for the two coupled relaxation oscillators in Fig. 10.5 with $V_0 = 3$, $R = C_1 = 1$, and $C_2 = 0.81$.

The equations that describe the operation are

$$\dot{V}_1 = (V_0 - V_1 - V_2)/RC_1 \quad \text{for lamp 1 not conducting}$$

$$V_1 = 0 \qquad\qquad\qquad\qquad \text{for lamp 1 conducting}$$

$$\dot{V}_2 = (V_0 - V_1 - V_2)/RC_2 \quad \text{for lamp 2 not conducting}$$

$$V_2 = 0 \qquad\qquad\qquad\qquad \text{for lamp 2 conducting.}$$

(10.3)

If the capacitors are identical, the neon lamps would fire simultaneously, but even the slightest difference will cause them to fire in succession. Even so, they tend to synchronize if the capacitors are similar. The voltage across each capacitor oscillates between zero and one, with only a slight difference in the period depending on the state of the other capacitor. Consequently, it is better to look at the total voltage $V = V_1 + V_2$ as a function of time as shown in Fig. 10.6. In this case C_2 is chosen as $0.81C_1$. The coefficient 0.81 is close to half the golden mean $(\sqrt{5} + 1)/2 = 1.6180339887\ldots$ to reduce the chance of frequency locking. (The golden mean is the 'most irrational' of the irrational numbers in the sense that it is the number most poorly approximated by a ratio of two small integers, or, equivalently, the irrational number whose representation as a continued fraction converges most slowly.) A variant of this circuit using tunnel diodes in place of the neon lamps was studied by Gollub *et al.* (1978).

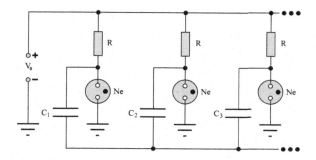

Fig. 10.7   Three of a possible N coupled relaxation oscillators.

10.4.2   *Many oscillators*

It is straightforward to couple more than two such oscillators by placing additional capacitors and neon lamps in series with the ones shown in Fig. 10.5. However, the voltage V_0 must be increased proportionally to the number of such oscillators to ensure that at least one of the capacitors achieves a value sufficient to fire its associated neon lamp.

A better circuit that can be extended to an arbitrary number of oscillators without needing to raise the source voltage is shown in Fig. 10.7. If the common point of the capacitors were grounded, the circuit would behave like N independent relaxation oscillators. However, the circuit as shown works rather differently. Due to asymmetries in the circuit, one of the neon lamps is the first to fire, dropping the voltage across the other lamps by the capacitive coupling. Unlike the previous circuits, the first lamp continues to conduct until a second lamp fires since the capacitors do not instantly discharge. The charging current for the capacitors whose lamps are not conducting discharges the capacitor connected to the lamp that is conducting, in fact driving its voltage negative, but not sufficiently so that its neon lamp immediately fires again.

The circuit is described by the equations

$$\dot{V_i} = \begin{cases} \left[V_0 - \sum_{k=1}^{N} (V_0 - V_k + V_j) \right] /RC_i & \text{for } i = j \\ (V_0 - V_i + V_j)/RC_i & \text{for } i \neq j \end{cases} \qquad (10.4)$$

where C_j is the capacitor associated with the neon lamp that is conducting. That lamp is extinguished, and the lamp associated with capacitor C_i begins conducting when its voltage V_i exceeds $1 - V_j$. A typical waveform for the voltage V_1 across capacitor C_1 for $N = 3$ is shown in Fig. 10.8.

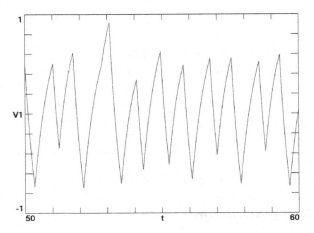

Fig. 10.8   Waveform for the capacitor voltage V_1 for the three coupled relaxation oscillators in Fig. 10.7 with $V_0 = 2$, $R = 1$, $C_1 = 1$, $C_2 = 2$, and $C_3 = 3$.

10.5   Forced Diode Resonator

A particularly simple chaotic circuit involves a sinusoidally forced series inductor and capacitor, one of which is nonlinear. One of the earliest such circuit using a varicap diode (or varactor) was studied by Linsay (1981). A reverse-biased ($V < 0$) varicap diode has a capacitance given approximately by $C = C_0/(1 - V/\Phi)^\gamma$, where typically $C_0 = 100$ picofarads, $\Phi = 0.6$ Volts, and $\gamma = 0.5$.

It turns out that the important nonlinearity occurs not from the region of reverse bias but rather from the very large capacitance that occurs when the diode is weakly forward biased, and thus an ordinary pn-junction diode will suffice as shown in Fig. 10.9. In fact, a silicon power diode (one designed for high current operation) tends to have more junction capacitance and thus allows a chaotic circuit to be constructed that operates at audio frequencies with a series inductance of less than 1 Henry so that the period-doubling bifurcations and chaos can be heard. This circuit has been studied by Testa *et al.* (1982), Rollins and Hunt (1982), and van Buskirk and Jeffries (1985), among others.

The equations that describe the operation of the circuit (van Buskirk and Jeffries, 1985) are

$$\dot{V} = (I - I_d)/C$$
$$\dot{I} = (V_0 \sin \Omega t - V)/L \tag{10.5}$$

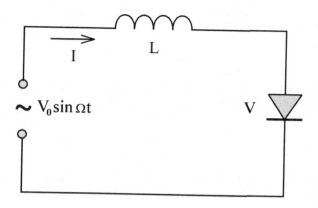

Fig. 10.9   Forced diode resonator.

where V is the forward diode voltage, I_d is the resistive current in the diode, assumed to be given by $I_d = I_0(e^{V/\Phi} - 1)$, and C is the diode junction capacitance, assumed to be given by $C = C_0 e^{V/\Phi}$ when the diode is forward biased in contrast to the previous case where it was reverse biased. The quantity Φ is approximately 26 millivolts at room temperature, and I_0 is the reverse bias current, which is typically very small ($\sim 10^{-12}$ Amperes) for a silicon diode and thus can be neglected. The exponentially rising junction capacitance has the same effect in limiting the forward voltage drop as does the conduction current, but it removes the dissipation from the system, which does not much effect its operation. Of course a real circuit would have some resistance R in the inductor and elsewhere, but if $Q = \Omega L/R$ is much greater than one, this resistance can be ignored.

Rather than try to choose parameters for a particular diode, we choose generic parameters that are elegant by the standards of this book. The voltage scale is set by the actual value of Φ, and the time scale is set by $\sqrt{LC_0}$ for the chosen components. A typical waveform for the diode voltage is shown in Fig. 10.10.

10.6   Saturating Inductor Circuit

Behavior similar to the forced diode resonator can be obtained by using a capacitor in place of the diode and allowing the inductor to be nonlinear. In particular, an inductor with an iron core will have an effective inductance that decreases with the magnitude of the current in the inductor due to

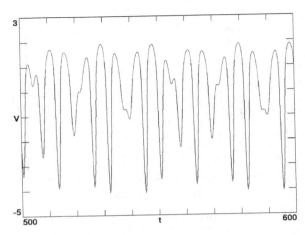

Fig. 10.10   Waveform for the diode voltage in the forced diode resonator in Fig. 10.9 with $V_0 = \Omega = L = \Phi = C_0 = 1$, $L = 0.3$, and $I_d = 0$, $\lambda = (0.0188, 0, -0.0188)$.

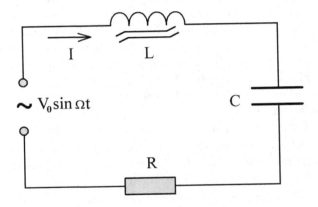

Fig. 10.11   Saturating inductor circuit.

saturation of the iron core. Therefore, a simple circuit as shown in Fig. 10.11 is expected to exhibit chaos.

The equations that describe the operation of the circuit are

$$\dot{V} = I/C$$
$$\dot{I} = (V_0 \sin \Omega t - V - IR)/L \tag{10.6}$$

where V is the voltage across the capacitor and the effective inductance is assumed to vary with current according to $L = L_0/(1 + I^2)$, which corresponds to a magnetic flux in the inductor that varies with current according

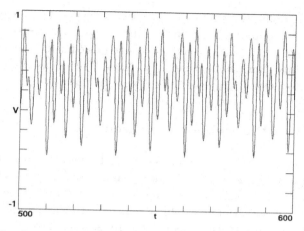

Fig. 10.12 Waveform for the capacitor voltage in the saturating inductor circuit in Fig. 10.11 with $C = V_0 = 1, \Omega = 0.1, L_0 = 0.2$, and $R = 0.02$, $\lambda = (0.0303, 0, -0.1807)$.

to $\Phi = \arctan I$. The equations exhibit chaos with $R = 0$, but a small resistance is included to be more realistic and to ensure that the system has an attractor that is not dependent on initial conditions, here taken as $V(0) = I(0) = 0$. A typical waveform for the capacitor voltage is shown in Fig. 10.12. If the inductor saturates at a different value of current, the forcing voltage must be changed proportionally, and the resulting capacitor voltage scales accordingly. Note that $Q = \Omega L_0 / R = 1$ for the chosen parameters, and so the inductor can be quite resistive without destroying the chaos.

Chua *et al.* (1982) numerically analyzed this circuit assuming a piecewise-linear variation of inductance with current and showed that chaotic solutions are expected. Dean (1994) experimentally observed chaos in the circuit when forced with a square wave rather than with a sine wave. His analysis also assumed a piecewise-linear variation of inductance with current and included hysteresis, which raises the dimension of the system by one. A more realistic, smoothly saturating inductor model similar to the one proposed here, but without hysteresis, was studied by Bartuccelli *et al.* (2007). It is safe to say that this simple circuit has yet to be fully explored, and the best model for its behavior is still lacking.

Forced oscillators such as this and the diode resonator in the previous section can be made into autonomous circuits by combining them with an active element such as an operational amplifier designed to produce a continuous periodic waveform, perhaps by placing the nonlinear element in the feedback loop of an amplifier to make it oscillate. However, a third reactive element (capacitor or inductor) is required since the system must be at least three-dimensional to exhibit chaos. The Wien-bridge oscillator, to be discussed shortly, provides an example of this idea.

10.7 Forced Piecewise-linear Circuit

A saturating operational amplifier can also be used to provide the nonlinearity in a forced LRC circuit, one example of which as suggested by Arulgnanam *et al.* (2009) is shown in Fig. 10.13 with the necessary power supplies for the operational amplifier omitted. In this circuit, the operational amplifier with its three associated resistors behaves like a negative resistance with a value of $-R$ provided the capacitor voltage is small enough not to saturate the amplifier, which is assumed to occur when its output voltage exceeds ± 1 Volt, corresponding to a capacitor voltage of $V = \pm 0.5$ Volts. Once the amplifier saturates, it behaves like a positive resistance with a value of approximately $R/3$ as if the three resistors were all in parallel with one another and in series with a voltage source of $\pm 2/3$ Volts. The actual values depend somewhat on how the amplifier behaves when it is driven to saturation, in particular whether the differential input resistance remains high or is clamped to a low value.

The equations that describe the operation of the circuit are

$$
\begin{aligned}
\dot{V} &= (I - g(V))/C \\
\dot{I} &= (V_0 \sin \Omega t - V)/L
\end{aligned}
\tag{10.7}
$$

where I is the current in the inductor and $g(V)$ is the current at the input of the operational amplifier circuit and is assumed to be given by $g(V) = 3V/R - 2[|V + 0.5| - |V - 0.5|]/R$ (Kennedy, 1992). A typical waveform for the capacitor voltage is shown in Fig. 10.14. If the operational amplifier saturates at a voltage higher than 1 Volt, the value of V_0 required to produce chaos is increased proportionally.

10.8 Chua's Circuit

Probably more has been written about Chua's circuit and its many variants than all other chaotic circuits combined. Good recent summaries of

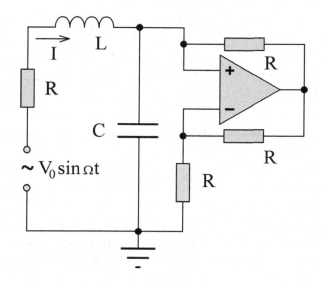

Fig. 10.13 Forced piecewise-linear circuit.

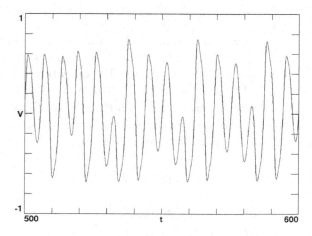

Fig. 10.14 Waveform for the capacitor voltage in the forced piecewise-linear circuit in Fig. 10.13 with $R = C = V_0 = \Omega = 1$, and $L = 2$, $\lambda = (0.0958, 0, -0.9346)$.

this large body of work are contained in the beautifully illustrated book by Bilotta and Pantano (2008) and in the recent book by Fortuna *et al.* (2009) commemorating the twenty-fifth anniversary of the invention of Chua's

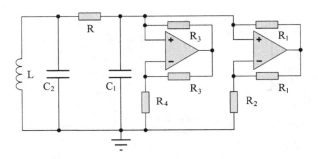

Fig. 10.15   Chua's circuit.

circuit in the fall of 1983 (Chua, 1992). The circuit was designed by Leon Chua and originally constructed to illustrate that the chaos modelled by the Lorenz equations is a robust physical phenomenon and not a numerical artifact.

By comparison with the previous examples and those to follow, the circuit is neither especially simple nor elegant, but it is included here because of its historical importance in the development and study of chaotic circuits and because it is capable of producing a huge variety of waveforms and strange attractors with a suitable choice of its seven parameters.

The basic circuit is shown in Fig. 10.15 with the necessary power supplies for the operational amplifiers omitted. The operational amplifiers are assumed identical and to saturate when their output reaches ± 1 Volt. A different saturation voltage has no effect on the circuit other than to proportionally change the values of all the voltages and currents.

The equations that describe the circuit are

$$\dot{V}_1 = [(V_2 - V_1)/R - I_a - I_b]/C_1$$
$$\dot{V}_2 = [(V_1 - V_2)/R + I]/C_2 \qquad (10.8)$$
$$\dot{I} = -V_2/L$$

where V_1 and V_2 are the voltages across the capacitors C_1 and C_2, respectively, I is the current in the inductor L, I_a is the current in R_2, and I_b is the current in R_4. The piecewise-linear nonlinearity arises from the saturating operational amplifiers that behave as negative resistances in their linear ranges and as positive resistances when saturated, with respective

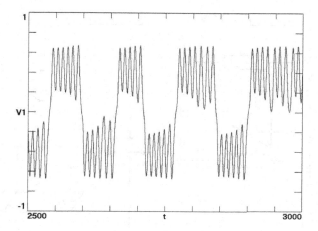

Fig. 10.16   Waveform for the voltage V_1 across the capacitor C_1 in Chua's circuit in Fig. 10.15 with $L = 0.5, R = 0.7, C_2 = 2, C_1 = 1, R_1 = 3, R_2 = 1, R_3 = 0.5$, and $R_4 = 1$ and initial conditions $(V_1, V_2, I) = (0.01, 0, 0)$, $\lambda = (0.0544, 0, -1.1263)$.

currents I_a and I_b given by

$$I_a = \begin{cases} -V_1/R_2 & \text{for } |V_1| \leq R_2/(R_1 + R_2) \\ (V_1 - \text{sgn}\, V_1)/R_1 & \text{for } |V_1| > R_2/(R_1 + R_2) \end{cases}$$

$$I_b = \begin{cases} -V_1/R_4 & \text{for } |V_1| \leq R_4/(R_3 + R_4) \\ (V_1 - \text{sgn}\, V_3)/R_3 & \text{for } |V_1| > R_4/(R_3 + R_4). \end{cases}$$

(10.9)

A typical waveform for the capacitor voltage V_1 with a simple set of parameters is shown in Fig. 10.16. Note that the initial condition $V_1 = V_2 = I = 0$ is an equilibrium point, and so V_1 is taken to have an initial value of 0.01 to start the oscillation. The resulting waveform resembles that for the x variable of the Lorenz system with its characteristic two-lobe structure.

10.9   Nishio's Circuit

An even simpler circuit in the spirit of Chua was proposed by Nishio *et al.* (1990) and studied in detail by Bonatto and Gallas (2008). It requires only a single operational amplifier operating in its linear range to produce a negative resistance, with the nonlinearity provided by a pair of parallel

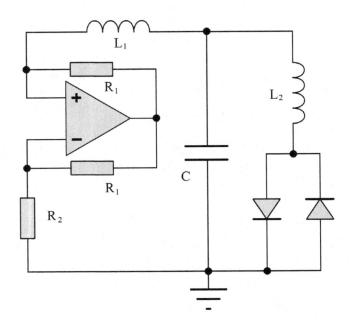

Fig. 10.17   Nishio's circuit.

back-to-back diodes as shown in Fig. 10.17. It has only nine components in contrast to Chua's circuit, which has twelve components.

The diodes are assumed to have a forward voltage drop of 1 Volt and to draw no current until the voltage reaches that value. A typical low-current silicon diode actually has a voltage drop of about 0.6 Volts, but high-power diodes have a voltage drop closer to 1 Volt at high current, and several such diodes can be placed in series to obtain a higher forward voltage if desired. Alternately, the circuit can be constructed with a pair of identical Zener diodes in series to obtain a larger voltage drop and a more abrupt onset of current when the critical voltage is reached. However, the voltage everywhere in the circuit, including the output of the operational amplifier, increases in proportion to the diode voltage, and so care must be taken to ensure that the operational amplifier does not saturate. Unlike the diode resonator circuit previously described, the diodes here are assumed to have negligible capacitance, or equivalently, that the circuit is designed to oscillate at a sufficiently low frequency that the diode capacitance is of no importance.

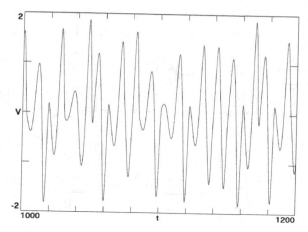

Fig. 10.18   Waveform for voltage V across capacitor C in Nishio's circuit in Fig. 10.17 with $C = R_1 = R_2 = L_2 = 1$ and $L_1 = 3$ with initial conditions $(V_1, V_2, I) = (0.01, 0, 0)$, $\lambda = 0.1332$.

The equations that describe the operation of the circuit are

$$\dot{V} = -(I_1 + I_2)/C$$
$$\dot{I}_1 = (V + I_1 R_2)/L_1 \qquad (10.10)$$
$$\dot{I}_2 = (V - \operatorname{sgn} I_2)/L_2.$$

As usual, the sgn I_2 term is replaced by $\tanh(200 I_2)$ for purposes of calculating the Lyapunov exponent and confirming the chaos. The factor of 200 is not critical, and any large value will suffice. This form is slightly simpler and more elegant as well as arguably more accurate than the approximation used by Nishio *et al.* (1990) and given by $0.5\gamma(|I_2 + 1/\gamma| - |I_2 - 1/\gamma|)$ with a typical value of $\gamma = 470$. Note that the equations do not depend on the resistance R_1, but it cannot be zero, and so it is taken arbitrarily as $R_1 = 1$. A typical waveform for the capacitor voltage is shown in Fig. 10.18.

10.10   Wien-bridge Oscillator

Circuits with inductors are problematic because inductors often behave in a nonideal manner. They have significant resistive losses, and both the resistance and inductance depend on frequency because of the skin effect. In addition, large inductors typically require an iron or ferrite core, which has its own complications due to saturation and hysteresis. Furthermore, it is

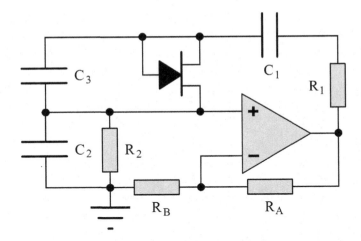

Fig. 10.19   Chaotic Wien-bridge oscillator.

difficult to fabricate integrated circuits with more than tiny values of inductance. Although an inductor can be replaced by a *gyrator circuit* consisting
of a capacitor, two resistors, and an operational amplifier (Antoniou, 1969;
Horowitz and Hill, 1989) and variants of Chua's circuit with only capacitors
have been built and studied (Morgül, 1995; Tôrres and Aguirre, 2000), it is
more sensible to design circuits from the outset with only capacitors. Since
autonomous chaotic systems require three first-order differential equations,
the corresponding circuit must contain three capacitors in addition to a
nonlinear element.

Wien-bridge oscillators are widely used to produce nearly sinusoidal
voltages without requiring inductors (Horowitz and Hill, 1989). With the
addition of an inductor or a third capacitor and a nonlinear element, they
can be used as building blocks for chaotic circuits (Elwakil and Soliman,
1997; Kiliç and Yildirim, 2008). An example of such a circuit with only nine
components including a single linear operational amplifier and no inductors
is the modified Wien-bridge oscillator (Namajūnas and Tamaševičius, 1995)
shown in Fig. 10.19. It requires the use of a junction field effect transistor (FET) that acts as a nonlinear voltage-controlled resistor with a small
resistance when the voltage across it is small, but with a nearly constant
current when the voltage exceeds some threshold, which is arbitrarily taken
as 1 Volt.

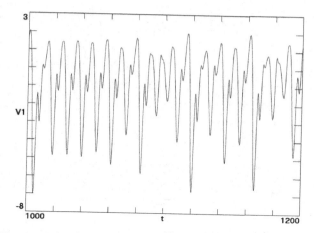

Fig. 10.20   Waveform for the voltage V_1 across capacitor C_1 in the chaotic Wien-bridge oscillator in Fig. 10.19 with $C_1 = C_2 = R_1 = R_2 = R_3 = R_B = 1, C_3 = 0.4$, and $R_A = 4$ with initial conditions $(V_1, V_2, V_3) = (0.01, 0, 0)$, $\lambda = (0.1066, 0, -2.1833)$.

The equations that describe the operation are

$$\dot{V}_1 = I_0/C_1$$
$$\dot{V}_2 = (I_0 - V_2)/C_2 \qquad (10.11)$$
$$\dot{V}_3 = [I_0 - I_3(V_3)]/C_3$$

where $I_0 = (R_A V_2/R_B - V_1 - V_3)/R_1$ is the current in the upper-right-hand branch. The current through the FET is assumed to be given by $I_3(V_3) = V_3/R_3$ for $V_3 \leq 1$ and $I_3(V_3) = 1$ for $V_3 > 1$. A typical waveform for the voltage $V_1(t)$ across capacitor C_1 is given in Fig. 10.20.

Inductorless Wien-bridge chaotic oscillators with as few as eight components including a single operational amplifier and a field effect transistor have been developed by Elwakil and Soliman (1997), but the equations that describe their operation are more complicated and depend upon a nonideal property of the operational amplifier, namely its gain–bandwidth product. A chaotic Wien-bridge oscillator can also be constructed with nine components, including one linear operational amplifier, one inductor, and a diode whose junction capacitance is essential to its operation (Elwakil and Kennedy, 1998), but its analysis is more difficult because four differential equations and a nonideal diode model are required. Because of the special nature of the nonlinear elements and the inelegant form of the equations that describe their operation, these systems will not be further discussed.

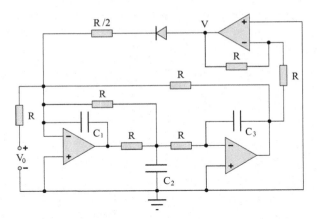

Fig. 10.21   Jerk circuit with absolute-value nonlinearity.

10.11   Jerk Circuits

A particularly simple design strategy for chaotic circuits is to start with one of the chaotic jerk systems described in Chapter 3 since the successive derivatives can be implemented with integrators, and design an analog computer to solve it. We will thus refer to such circuits as 'jerk circuits' (Sprott, 2000a,b). Any of the systems in Table 3.3 are good candidates for such an implementation.

10.11.1   *Absolute-value case*

The system MO_0 in Table 3.3 involves a nonlinearity of the form $|x|$, which can be easily implemented with a diode since it is basically a half-wave rectifier circuit (converting an ac voltage into an unfiltered dc voltage). One circuit that performs the required operation is shown in Fig. 10.21. In this circuit, one of the integrations (the one that involves C_2) is performed passively (using only an RC without the accompanying operational amplifier), while the other two are performed actively with operational amplifiers, but this poses no difficulty because the passive integration simply adds some dissipation not present with the active integrators, which is required in any case.

The equation that governs the behavior of the circuit, assuming all resistors have the same value of $R = 1$ and the diode is ideal (no forward

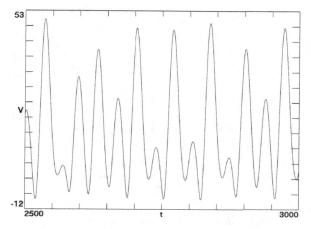

Fig. 10.22 Waveform for the voltage V in the jerk circuit with an absolute-value non-linearity in Fig. 10.21 with $V_0 = C_1 = C_3 = 1$ and $C_2 = 30$, $\lambda = (0.0132, 0, -0.1132)$.

voltage drop and no junction capacitance), is

$$\dddot{V} + A\ddot{V} + B\dot{V} + CV = D|V| - E \qquad (10.12)$$

where $A = 3/C_2, B = 1/C_1C_2, C = D = 1/2C_1C_2C_3$, and $E = V_0/C_1C_2C_3$. The voltage waveform $V(t)$, which is also the voltage across the capacitor C_3, is shown in Fig. 10.22. Note that the voltage $V(t)$ is relatively large compared to V_0, and thus the operational amplifiers must be capable of handling this voltage without saturating, or else a smaller value of V_0 should be used. With only a slight increase in complexity (one additional diode), the third operational amplifier and diode can be replaced with a precision (active) inverting half-wave rectifier (Horowitz and Hill, 1989), which compensates for the nonideal properties (especially the forward voltage drop) of the diode. A version of the circuit using this improvement has been studied by Kiers *et al.* (2004a). What the circuit lacks in elegance, it makes up for by the simplicity of the equation that describes its operation.

10.11.2 *Single-knee case*

With a slight simplification of the circuit in Fig. 10.21, the circuit can be used to solve the model MO_1 in Table 3.3, which has a nonlinearity of the form $\max(x, 0)$. The modified circuit is shown in Fig. 10.23.

The equation that governs the behavior of the circuit, assuming all resistors have the same value of $R = 1$ and the diode is ideal (no forward

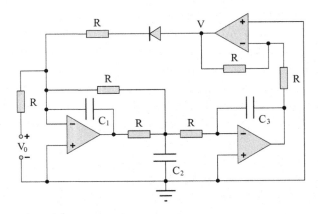

Fig. 10.23   Jerk circuit with a single-knee nonlinearity.

voltage drop and no junction capacitance), is

$$\ddot{V} + A\ddot{V} + B\dot{V} + CV = D\max(V,0) - E \tag{10.13}$$

where $A = 3/C_2, B = 1/C_1C_2, C = D = 1/2C_1C_2C_3$, and $E = V_0/C_1C_2C_3$. The voltage waveform $V(t)$, which is also the voltage across the capacitor C_3, is shown in Fig. 10.24. The voltage $V(t)$ is even larger than in the previous circuit, and so special care must be taken to avoid saturating the operational amplifiers. A slightly more complicated version of this circuit has been studied by Kiers *et al.* (2004b).

10.11.3   *Signum case*

A particularly simple jerk circuit uses the saturating property of an operational amplifier to implement the signum function $\operatorname{sgn} V$, which takes on a value of $+1$ or -1 depending on whether V is positive or negative, respectively. Such a circuit corresponding to model MO_2 of Table 3.3 is shown in Fig. 10.25 (Sprott, 2000b). The upper right operational amplifier has no feedback and thus acts as a comparator circuit performing the signum operation. Its output can be scaled appropriately by adjusting the resistor at its output, but we will assume the output voltage saturates at ± 1 Volt and take all the resistors to have equal values of $R = 1$. Because of the attenuation in the two passive integrators, the left-hand operational amplifier must have a high gain, and so care must be taken to ensure that it does not saturate, perhaps by reducing the output of the comparator. The voltage at its output is given by $V_1 = C_2C_3\ddot{V} + (2C_2 + 2C_3)\dot{V} + 3V$

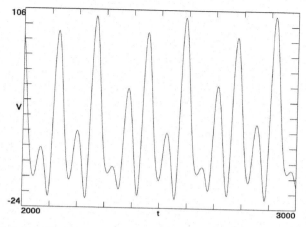

Fig. 10.24 Waveform for the voltage V in the jerk circuit with a single-knee nonlinearity in Fig. 10.23 with $V_0 = 1, C_1 = 2, C_2 = 60$ and $C_3 = 1$, $\lambda = (0.0065, 0, -0.0565)$.

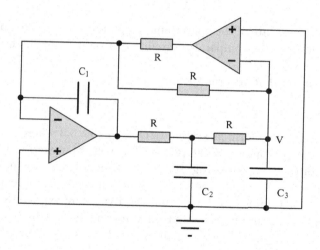

Fig. 10.25 Jerk circuit with signum nonlinearity.

and has a maximum value of about 9.6 Volts when the comparator delivers an output of ± 1 Volt for the capacitor values given below.

This circuit (tested, but not previously published) with nine components is considerably simpler than Chua's circuit and comparable to Nishio's circuit but with no inductors. There is one active integrator and two passive integrators. Its operation is limited to relatively low frequencies as deter-

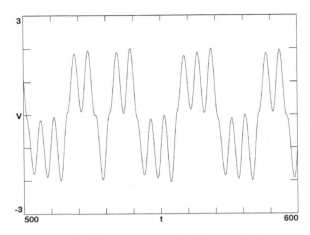

Fig. 10.26 Waveform for the voltage V across capacitor C_3 in the jerk circuit with a signum nonlinearity in Fig. 10.25 with $C_1 = 0.1, C_2 = 1, C_3 = 2$ and initial conditions $(V, \dot{V}, \ddot{V}) = (0.01, 0, 0), \lambda = (0.0587, 0, -3.0587)$.

mined by the slew rate of the operational amplifiers, but special operational amplifiers designed as comparator circuits can be used for high-frequency operation.

The equation that governs the circuit is

$$\dddot{V} + A\ddot{V} + B\dot{V} = C(\text{sgn}\,V - V), \qquad (10.14)$$

which is of the same form as system MO_2 in Table 3.3 where $A = 2/C_2 + 2/C_3, B = 3/C_2C_3$, and $C = 1/C_1C_2C_3$. The voltage waveform $V(t)$ across the capacitor C_3 as shown in Fig. 10.26 is a two-lobe attractor similar to the Lorenz attractor. As usual, the signum function sgn V is approximated by $\tanh(200V)$ for purposes of estimating the Lyapunov exponents. A much more complicated circuit using sixteen components with five operational amplifiers that solves the same equation has been studied by Elwakil and Kennedy (2001).

10.11.4 Signum variant

If one allows inductors, the circuit in Fig. 10.24 can be further simplified by replacing the two passive RC integrators with a single LC circuit as shown in Fig. 10.27. This circuit (tested, but not previously published) with only seven components is simpler than Chua's circuit and Nishio's circuit, both of which also have inductors.

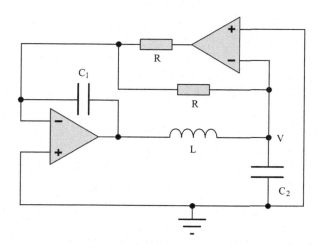

Fig. 10.27   Jerk circuit variant with signum nonlinearity.

This circuit is governed by the same equation as the previous circuit,

$$\dddot{V} + A\ddot{V} + B\dot{V} = C(\operatorname{sgn} V - V), \qquad (10.15)$$

except that the parameters are given by $A = 1/C_2, B = 1/LC_2$, and $C = 1/LC_1C_2$. As before, the resistors are assumed to have a value of $R = 1$, and the comparator is assumed to produce an output voltage of ± 1 Volts. The resistor at the output of the comparator can be increased proportionately if the comparator produces a larger output voltage. The waveform $V(t)$ across the capacitor C_2 is shown in Fig. 10.28. For the parameters given, the output of the operational amplifier that serves as an active integrator reaches a value of 12 Volts. Practical values for a circuit that oscillates in the audible range (~ 1 kHz) are given by $R = 1$ kilohm, $C_1 = 0.5$ microfarad, $C_2 = 1$ microfarad, and $L = 1$ Henry. As usual, the signum function $\operatorname{sgn} V$ is approximated by $\tanh(200V)$ for purposes of estimating the Lyapunov exponents.

10.12   Master–slave Oscillator

Chapter 6 included many examples of master–slave oscillators in which one oscillator provides an input for the second oscillator but is not affected by it. Such a system can be easily implemented electronically, one example of which (untested) using signum nonlinearities is shown in Fig. 10.29. The

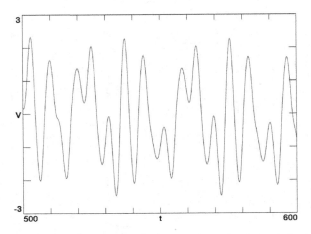

Fig. 10.28 Waveform for the voltage V across capacitor C_2 in the jerk circuit variant with a signum nonlinearity in Fig. 10.27 with $C_1 = 0.5, C_2 = 1, L = 1$ and initial conditions $(V, \dot{V}, \ddot{V}) = (0.01, 0, 0)$, $\lambda = (0.1535, 0, -1.1535)$.

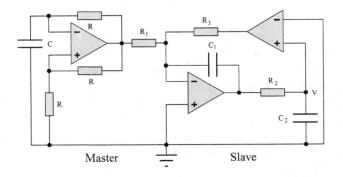

Fig. 10.29   Master–slave oscillator.

master relaxation oscillator on the left produces a square wave output with a frequency of $\Omega = \pi/RC \ln(3)$ with an amplitude assumed to be ± 1 Volt, and the two operational amplifiers on the right constitute the signum slave oscillator.

The voltage V across the capacitor C_2 is governed by the equation

$$\ddot{V} + A\dot{V} + B \operatorname{sgn} V = C \operatorname{sgn}(\sin \Omega t) \tag{10.16}$$

where the parameters are given by $A = 1/R_2C_2, B = 1/R_2R_3C_1C_2$, and $C = 1/R_1R_2C_1C_2$. The resistors and capacitors without subscripts are

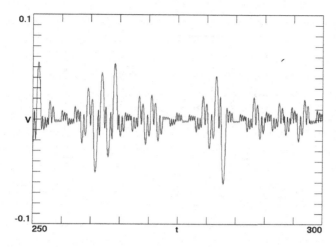

Fig. 10.30 Waveform for the voltage V across capacitor C_2 in the master–slave oscillator in Fig. 10.29 with $R = R_1 = R_2 = C = C_1 = C_2 = 1, R_3 = 0.6$ and initial conditions $(V, \dot{V}, t) = (0, 0, 0)$, $\lambda = (0.9121, 0, -1.9121)$.

assumed to have values of $R = C = 1$, and the comparator is assumed to produce an output voltage of ± 1 Volts. The waveform $V(t)$ across the capacitor C_2 is shown in Fig. 10.30.

10.13   Ring of Oscillators

Chapter 7 included many examples of systems consisting of a ring of elements that by themselves are not chaotic, but that become so when connected symmetrically into a ring. One such circuit (untested) that follows the previous examples with a signum nonlinearity is shown in Fig. 10.31. The upper operational amplifier serves as a comparator and is assumed to saturate when its output voltage reaches ± 1 Volt.

This system solves the equation

$$\ddot{V}_i + A\dot{V}_i + B(V_i - \operatorname{sgn} V_i) = -CV_{i-1} \tag{10.17}$$

where $A = (2/R + 1/R_k)/C_2, B = 1/R^2C_1C_2$, and $C = 1/R_kRC_1C_2$. The predicted waveform for one of the oscillators in a ring of eleven is shown in Fig. 10.32. The number $N = 11$ was chosen because it is a prime number that is not too small but that makes construction manageable. Other values of N should also work but will require some adjustment of the

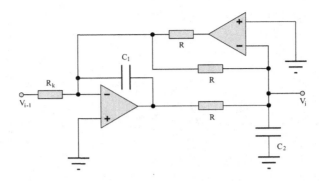

Fig. 10.31   One of N oscillators in a ring with a signum nonlinearity.

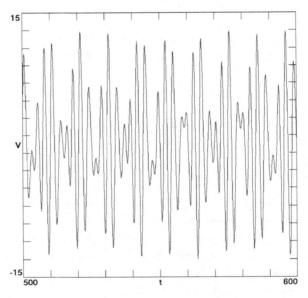

Fig. 10.32   Waveform for the output voltage of one of the eleven oscillators in a ring as in Fig. 10.31 with $C_1 = 0.1, C_2 = 1.3, R_k = 1.5, R = 1$ and initial conditions $V_i = (0.01, 0, 0, \ldots), \lambda = 0.1275$.

component values to achieve chaotic oscillations. For verifying the chaos and estimating the Lyapunov exponent, the $\operatorname{sgn} V_i$ term is replaced with $\tanh(500 V_i)$. Note that each element in the ring has only two capacitors and so, by itself, cannot exhibit chaos since the governing equation is only second order.

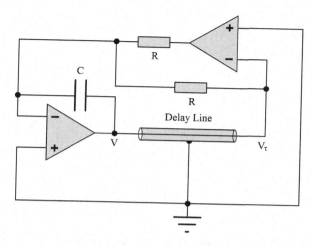

Fig. 10.33   Delay-line oscillator with signum nonlinearity.

10.14   Delay-line Oscillator

Chapter 9 dealt with chaotic systems governed by delay differential equations. These systems can be implemented electronically using a delay line, which can be as simple as a long coaxial cable. Such a delay line can be considered as a distributed series inductance and parallel capacitance as shown in Fig. 10.33 (untested), which bears a striking resemblance to Fig. 10.27. The single LC circuit that provides a frequency-dependent phase shift in Fig. 10.27 has been replaced with a distributed device producing a fixed time delay for all frequencies. As before, the operational amplifier in the upper right acts as a comparator, returning an output of ± 1 Volt, depending on the polarity of its input signal.

With $C = R = 1$, the circuit exactly solves the equation

$$\dot{V} = \operatorname{sgn} V_\tau - V_\tau, \tag{10.18}$$

which is identical to Eq. (9.7) and thus has the same solution shown as a time series $V(t)$ in Fig. 10.34. For chaos, the time delay is set at $\tau = 2$, which in real units means a time in seconds equal to twice RC in Ohm-Farads. To avoid reflections and resistive losses, the delay line should have a characteristic impedance equal to R and a resistance much less than R.

Other more complicated circuits with time delays have been proposed and tested. For example, Namajūnas *et al.* (1995) constructed such a circuit

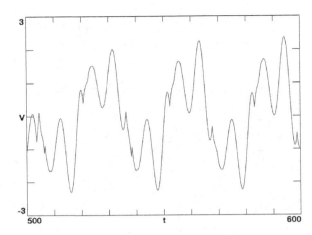

Fig. 10.34 Waveform for the voltage V in the delay-line oscillator with a signum non-linearity in Fig. 10.33 with $C = R = 1$, $\tau = 2$ and initial condition $V_0 = 0.1$, $\lambda = 0.0656$.

to solve the Mackey–Glass equation in the previous chapter, Lu and He (1996) used the same technique for a five-segment piecewise-linear system, Tamaševičius *et al.* (2006) studied a similar variant with a three-segment piecewise-linear system, and Wang *et al.* (2001) studied a time-delayed version of Chua's circuit. None of these systems is as simple as the one proposed here.

The study of chaotic circuits has blossomed in recent years and will continue to be an active area of research, but it is unlikely that circuits much simpler than those shown in this chapter will be developed since the conditions necessary for chaos in circuits is similar to those necessary in the differential equations that describe them. It is truly remarkable that chaos, which was largely unknown a few short decades ago, can now be observed in simple equations and electrical circuits that are not only potentially useful but are also strikingly elegant.

Bibliography

Ablowitz, M. J. (1991). *Solitons, Nonlinear Evolution Equations and Inverse Scattering* (Cambridge University Press, Cambridge).

Abraham, R. and Ueda, Y. (2000). *The Chaos Avant-Garde: Memories of the Early Days of Chaos Theory* (World Scientific, Singapore).

Abramowitz, M. and Stegun, I. A. (1972). *Handbook of Mathematical Functions with Formulas, Graphs, and Mathematical Tables* (Dover, New York).

Afraimovich, V. H. and Lin, W. W. (1998). Synchronization in lattices of coupled oscillators with Neumann/periodic boundary conditions, *Dynam. Stabil. Syst.* **13**, pp. 237–264.

Aguirre, J., Viana, R. L., and Sanjuán, M. A. F. (2009). Fractal structures in nonlinear dynamics, *Rev. Mod. Phys.* **81**, pp. 333–386.

Ahmad, W. M. (2006). A simple multi-scroll hyperchaotic system, *Chaos, Solitons & Fractals* **27**, pp. 1213–1219.

Allan, D. W. (1962). On the behaviour of systems of coupled dynamos, *Math. Proc. Camb. Phil. Soc.* **58**, pp. 671–693.

Alvarez-Ramirez, J., Delgado-Fernandez, J. and Espinosa-Paredes, G. (2005). The origin of a continuous two-dimensional "chaotic" dynamics, *Int. J. Bifurcat. Chaos Appl. Sci. Eng.* **15**, pp. 3023–3029.

Amari, S., Yoshida, K., and Kanatani, K. (1977). A mathematical foundation for statistical neurodynamics, *SIAM J. Appl. Math.* **33**, pp. 95–126.

an der Heiden, U. and Mackey, M. C. (1982). The dynamics of production and destruction: Analytic insight into complex behavior, *J. Math. Biol.* **16**, pp. 75–101.

Anishchenko, V. S. and Strelkova, G. I. (1997). Attractors of dynamical systems. In *Control of Oscillations and Chaos, Proceedings, 1997 International Conference*, Vol. 3, pp. 498–503.

Antoniou, A. (1969). Realisation of gyrators using OpAmps and their use in RC-active network synthesis, *Proc. IEE* **116**, pp. 1838–1850.

Arneodo, A., Coullet, P., and Tresser, C. (1980). Occurence of strange attractors in three-dimensional Volterra equations, *Phys. Lett. A* **79**, pp. 259–263.

Arneodo, A., Coullet, P., and Tresser, C. (1981). A possible mechanism for the onset of turbulence, *Phys. Lett. A* **81**, pp. 197–201.

Arneodo, A., Coullet, P., and Tresser, C. (1982). Oscillators with chaotic behavior: An illustration of a theorem by Shilnikov, *J. Stat. Phys.* **27**, pp. 171–182.

Arnold, V. I. (1978). *Mathematical Methods of Classical Mechanics* (Springer, New York).

Arulgnanam, A., Thamilmaran, K., and Daniel, M. (2009). Chaotic dynamics with high complexity in a simplified new nonautonomous nonlinear electronic circuit, *Chaos, Solitons & Fractals* **42**, pp. 2246–2253.

Auvergne, M. and Baglin, A. (1985). A dynamical instability as a driving mechanism for stellar oscillations, *Astron. Astrophys.* **142**, pp. 388–392.

Baker, G. L. and Blackburn, J. A. (2005). *The Pendulum: A Case Study in Physics* (Oxford University Press, Oxford).

Baker, G. L. and Gollub, J. P. (1996). *Chaotic Dynamics: An Introduction* (2nd edn) (Cambridge University Press, Cambridge).

Bartuccelli, M. V., Deane, J. H. B., and Gentile, G. (2007). Bifurcation phenomena and attractive periodic solutions in the saturating inductor circuit, *Proc. Math. Phys. Eng. Sci.* **463**, pp. 2351–2369.

Battelino, P. M., Grebogi, C., Ott, E., and Yorke, J. A. (1988). Multiple coexisting attractors, basin boundaries and basic sets, *Phys. Nonlinear Phenom.* **32**, pp. 296–305.

Becker, T. and Weispfenning, V. (1993). *Gröbner Bases: A Computational Approach to Commutative Algebra* (Springer, New York).

Bellen, A. and Zennaro, M. (2003). *Numerical Methods for Delay Differential Equations* (Oxford University Press, Oxford).

Bernhardt, P. A. (1991). The autonomous chaotic relaxation oscillator: An electrical analogue to the dripping faucet, *Phys. Nonlinear Phenom.* **52**, pp. 489–527.

Bilotta, E. and Pantano, P. (2008). *A Gallery of Chua Attractors* (World Scientific, Singapore).

Birkhoff, G. and Rota, G. C. (1978). *Ordinary Differential Equations* (3rd edn) (Wiley, New York), pp. 134.

Bonatto, C. and Gallas, J. A. C. (2008). Periodicity hub and nested spirals in the phase diagram of a simple resistive circuit, *Phys. Rev. Lett.* **101**, pp. 054101-1–4.

Bonatto, C., Gallas, J. A. C., and Ueda, Y. (2008). Chaotic phase similarities and recurrences in a damped-driven Duffing oscillator, *Phys. Rev. E* **77**, pp. 026217-1–5.

Bondi, H. (1957). Negative mass in general relativity, *Rev. Mod. Phys.* **29**, pp. 423–428.

Bonhoeffer, K. F. (1953). Modelle der Nervenerregung, *Naturwissenschaften* **40**, pp. 301–311.

Brayton, R. K. (1966). Bifurcation of periodic solutions in a nonlinear difference-differential equation of neutral type, *Q. Appl. Math.* **24**, 215–224.

Brown, R. (1993). From Chua's circuit to the generalized Chua map, *J. Circ. Syst. Comput.* **3**, pp. 11–32.

Brummitt, C. D. and Sprott, J. C. (2009). A search for the simplest chaotic partial differential equation, *Phys. Lett. A* **373**, pp. 2717–2721.

Bullard, E. (1955). The stability of a homopolar dynamo, *Math. Proc. Camb. Phil. Soc.* **51**, pp. 744–760.

Cafagna, D. and Grassi, G. (2003). Hyperchaotic coupled Chua circuits: An approach for generating new $n \times m$-scroll attractors, *Int. J. Bifurcat. Chaos Appl. Sci. Eng.* **13**, pp. 2537–2550.

Camacho, E., Rand, R., and Howland, H. (2004). Dynamics of two van der Pol oscillators coupled via a bath, *Int. J. Solid. Struct.* **41**, pp. 2133–2143.

Canuto, C., Hussani, M. Y., Quarteroni, A., and Zang, T. A. (2006). *Spectral Methods: Fundamentals in Single Domains* (Springer-Verlag, Berlin).

Cartwright, M. L. and Littlewood, J. E. (1945). On nonlinear differential equation of the second order. I. The equation $\ddot{x} - k(1 - y^2)\dot{x} + y = b\lambda k \cos(\lambda t + \alpha)$, k large, *J. Lond. Math. Soc.* **20**, pp. 180–189.

Cartwright, J. H. E., Hernández-Garcia, E., and Piro, O. (1997). Burridge-Knopoff models as elastic excitable media, *Phys. Rev. Lett.* **79**, pp. 527–530.

Case, W. B. (1996). The pumping of a swing from the standing position, *Am. J. Phys.* **64**, pp. 215–220.

Case, W. B. and Swanson, M. A. (1990). The pumping of a swing from the seated position, *Am. J. Phys.* **58**, pp. 463–467.

Chen, F. F. (1984). *Introduction to Plasma Physics and Controlled Fusion* (2nd edn) (Plenum Press, New York).

Chen, G. and Ueta, T. (1999). Yet another chaotic attractor, *Int. J. Bifurcat. Chaos Appl. Sci. Eng.* **9**, pp. 1465–1466.

Chen, Z., Yang, Y., Qi, G., and Yuan, Z. (2007). A novel hyperchaos system only with one equilibrium, *Phys. Lett. A* **360**, pp. 696–701.

Chirikov, B. V. (1979). A universal instability of many-dimensional oscillator systems, *Phys. Rep.* **52**, pp. 263–379.

Chiu, C. H., Lin, W. W., and Wang, C. S. (2001). Synchronization in lattices of coupled oscillators with various boundary conditions, *Nonlinear Anal. Theor. Meth. Appl.* **46**, pp. 213–229.

Chlouverakis, K. E. and Sprott, J. C. (2006). Chaotic hyperjerk systems, *Chaos, Solitons & Fractals* **28**, pp. 739–746.

Chlouverakis, K. E. and Sprott, J. C. (2007). Hyperlabyrinth chaos: From chaotic walks to spatiotemporal chaos, *Chaos* **17**, pp. 023110-1–8.

Chua, L. O. (1992). The genesis of Chua's circuit, *Archiv für Elektronik und Übertragungstechnik* **46**, pp. 250–257.

Chua, L. O. (1994). Chua's circuit: Ten years later, *IEICE Trans. Fund. Electron. Comm. Comput. Sci.* **E77-A**, pp. 1811–1822.

Chua, L. O., Hasler, M., Neirynck, J., and Verburgh, P. (1982). Dynamics of a piecewise-linear resonant circuit, *IEEE Trans. Circ. Syst.* **29**, pp. 535–547.

Chua, L. O., Komuro, M., and Matsumoto, T. (1986). The double scroll family, *IEEE Trans. Circ. Syst.* **33**, pp. 1073–1118.

Cook, A. E. and Roberts, P. H. (1970). The Rikitake two disk dynamo system, *Math. Proc. Camb. Phil. Soc.* **68**, pp. 547–569.

Coullet, P., Tresser, C., and Arnéodo, A. (1979). Transition to stochasticity for a class of forced oscillators, *Phys. Lett. A* **72**, pp. 268–270.

Cryer, C. W. (1972). Numerical methods for functional differential equations. In *Delay and Functional Differential Equations and their Applications* (ed. K. Schmitt), pp. 17–101 (Academic Press, New York).

Cummings, F. W., Dixon, D. D., and Kaus, P. E. (1992). Dynamical model of the magnetic field of neutron stars, *Astrophys. J.* **386**, pp. 215–221.

Darrigol, O. (2003). The spirited horse, the engineer, and the mathematician: Water waves in nineteenth-century hydrodynamics, *Archive for History of Exact Sciences* **58**, pp. 21–95.

Dean, J. H. B. (1994). Modeling the dynamics of nonlinear inductor circuits, *IEEE Trans. Magn.* **30**, pp. 2795–2801.

Den Hartog, J. P. (1930). Forced vibrations with combined viscous and Coulomb damping, *Phil. Mag.* **9**, pp. 801–817.

Ditto, W. L., Spano, M. L., Savage, H. T., Rauseo, S. N., Heagy, J., and Ott, E. (1990). Experimental observation of a strange nonchaotic attractor, *Phys. Rev. Lett.* **65**, pp. 533–536.

Dixon, D. D., Cummings, F. W., and Kaus, P. E. (1993). Continuous "chaotic" dynamics in two dimensions, *Phys. Nonlinear Phenom* **65**, pp. 109–116.

Drazin, P. G. and Johnson, R. S. (1989). *Solitons: An Introduction* (Cambridge University Press, Cambridge).

Driver, R. D. (1984). A neutral system with state dependent delay, *J. Differ. Equat.* **54**, pp. 73–86.

Duan, Z., Wang, J.-Z., and Huang, L. (2005). Attraction/repulsion functions in a new class of chaotic systems, *Phys. Lett. A* **335**, pp. 139–149.

Duffing, G. (1918). *Erzwungene Schwingungen bei Veränderlicher Eigenfrequenz und ihre Technische Bedeutung, Sammlung Vieweg, Heft 41/42* (Vieweg, Braunschweig).

Earnshaw, S. (1842). On the nature of the molecular forces which regulate the constitution of the luminiferous ether, *Trans. Camb. Phil. Soc.* **7**, pp. 97–112.

Eichhorn, R., Linz, S. J., and Hänggi, P. (1998). Transformations of nonlinear dynamical systems to jerky motion and its application to minimal chaotic flows, *Phys. Rev. E* **58**, pp. 7151–7164.

Eichhorn, R., Linz, S. J., and Hänggi, P. (2002). Simple polynomial classes of chaotic jerky dynamics, *Chaos, Solitons & Fractals* **13**, pp. 1–15.

Elwakil, A. S. and Kennedy, M. P. (1998). High frequency Wien-type chaotic oscillator, *Electron. Lett.* **34**, pp. 1161–1162.

Elwakil, A. S. and Kennedy, M. P. (2001). Construction of classes of circuit-independent chaotic oscillators using passive-only nonlinear devices, *IEEE Trans. Circ. Syst.* **48**, pp. 289–307.

Elwakil, A. S. and Solimon, A. M. (1997). A family of Wien type oscillators modified for chaos, *Int. J. Circuit Theory Appl.* **25**, pp. 561–579.

Elwakil, A. S., Salama, K. N., and Kennedy, M. P. (2000). A system for chaos generations and its implementation in monolithic form, *IEEE Trans. Circ. Syst.* **47**, pp. 217–220.

Elwakil, A. S., Özoğuz, S., and Kennedy, M. P. (2002). Creation of a complex butterfly attractor using a novel Lorenz-type system, *IEEE Trans. Circ. Syst.* **49**, pp. 527–530.

Erneux, T. (2009). *Applied Delay Differential Equations*, Surveys and Tutorials in the Applied Mathematical Sciences (Springer, New York).

Farmer, J. D. (1982). Chaotic attractors of an infinite dimensional system, *Phys. Nonlinear Phenom.* **4**, pp. 366–393.

Farmer, J. D. (1985). Sensitive dependence on parameters in nonlinear dynamics, *Phys. Rev. Lett.* **55**, pp. 351–354.

Feeny, B. (1992). A nonsmooth Coulomb friction oscillator, *Phys. Nonlinear Phenom.* **59**, pp. 25–38.

Feeny, B. and Moon, F. C. (1994). Chaos in a forced dry-friction oscillator: Experiments and numerical modeling, *J. Sound Vib.* **170**, pp. 303–323.

Feudel, U. and Grebogi, C. (2003). Why are chaotic attractors rare in multistable systems?, *Phys. Rev. Lett.* **91**, pp. 134102-1–4.

Feudel, U., Grebogi, C., Hunt, B. R., and Yorke, J. A. (1996). Map with more than 100 coexisting low-period periodic attractors, *Phys. Rev. E* **54**, pp. 71–81.

Feudel, U., Kuznetsov, S., and Pikovsky, A. (2006). *Strange Nonchaotic Attractors* (World Scientific, Singapore).

FitzHugh, R. (1960). Thresholds and plateaus in the hodkin-huxley nerve equations, *J. Gen. Physiol.* **43**, pp. 867–896.

FitzHugh, R. (1961). Impulses and physiological states in theoretical models of nerve membrane, *Biophys. J.* **1**, pp. 445–446.

Flaherty, J. E. and Hoppensteadt, F. C. (1978). Frequency entrainment of a forced van der Pol oscillator, *Stud. Appl. Math.* **58**, pp. 500–544.

Fortuna, L., Frasca, M., and Xibilia, M. G. (2009). *Chua's Circuit Implementations: Yesterday, Today and Tomorrow* (World Scientific, Singapore).

Funahashi, K. and Nakamura, Y. (1993). Approximation of dynamical systems by continuous time recurrent neural networks, *Neural Networks* **6**, pp. 801–806.

Gans, R. F. (1995). When is cutting chaotic?, *J. Sound Vib.* **188**, pp. 75–83.

Gao, T., Chen, G., Chen, Z., and Cang, S. (2007). The generation and circuit implementation of a new hyper-chaos based upon Lorenz system, *Phys. Lett. A* **361**, pp. 78–86.

Gear, C. W. (1971). *Numerical Initial Value Problems in Ordinary Differential Equations* (Prentice Hall, Englewood Cliffs, NJ).

Geist, K., Parlitz, U., and Lauterborn, W. (1990). Comparison of different methods for computing Lyapunov exponents, *Prog. Theor. Phys.* **83**, pp. 875–893.

Gerjuoy, E. (1949). On Newton's third law and the conservation of momentum, *Am. J. Phys.* **17**, pp. 477–482.

Gilmore, R. and Letellier, C. (2007). *The Symmetry of Chaos* (Oxford University Press, Oxford).

Gleick, J. (1987). *Chaos: Making a New Science* (Viking, New York).

Gollub, J. P., Brunner, T. O., and Danly, B. G. (1978). Periodicity and chaos in coupled nonlinear oscillators, *Science* **200**, pp. 48–50.

González-Miranda, J. M. (2006). Highly incoherent phase dynamics in the Sprott E chaotic flow, *Phys. Lett. A* **352**, pp. 83–88.

Gottlieb, H. P. W. (1996). Question 38. What is the simplest jerk function that gives chaos?, *Am. J. Phys.* **64**, pp. 525.

Gottlieb, H. P. W. and Sprott, J. C. (2001). Simplest driven conservative chaotic oscillator, *Phys. Lett. A* **291**, pp. 385–388.

Grantham, W. J. and Lee, B. (1993). A chaotic limit cycle paradox, *Dynam. Contr.* **3**, pp. 159–173.

Grassberger, P. and Procaccia, I. (1984). Dimensions and entropies of strange attractors from fluctuating dynamics approach, *Phys. Nonlinear Phenom.* **13**, pp. 34–54.

Grebogi, C., Ott, E. Pelikan, S., and Yorke J. A. (1984). Strange attractors that are not chaotic, *Phys. Nonlinear Phenom.* **13**, pp. 261–268.

Grebogi, C., McDonald, S. W., Ott, E., and Yorke, J. A. (1985). Exterior dimension of fat fractals, *Phys. Lett. A* **110**, pp. 1–4.

Gutzwiller, M. C. (1998). Moon-Earth-Sun: The oldest three-body problem, *Rev. Mod. Phys.* **70**, pp. 589–639.

Gyergyek, T., Čerček, M., and Stanojević, M. (1997). Experimental evidence of periodic pulling in a weakly magnetized discharge plasma column, *Contrib. Plasma Phys.* **37**, pp. 399–416.

Hale, J. K. (1977). *Theory of Functional Differential Equations* (Springer, New York).

Hayashi, C., Ueda, Y., Akamatsu, N., and Itakura, H. (1970). On the behavior of self-oscillatory systems with external force (in Japanese), *Trans. Inst. Electron. Comm. Eng. Jpn.* **53-A**, pp. 150–158.

Haykin, S. (1999). *Neural Networks – A Comprehensive Foundation* (2nd edn) (Prentice Hall, Upper Saddle River, NJ).

Heidel, J. and Zhang, F. (1999). Nonchaotic behaviour in three-dimensional quadratic systems II. The conservative case, *Nonlinearity* **12**, pp. 617–633.

Heidel, J. and Zhang, F. (2007). Nonchaotic and chaotic behavior in three-dimensional quadratic systems: Five-one conservative cases, *Int. J. Bifurcat. Chaos Appl. Sci. Eng.* **17**, pp. 2049–2072.

Hénon, M. (1982). On the numerical computation of Poincaré maps, *Phys. Nonlinear Phenom.* **5**, pp. 412–414.

Hénon, M. and Heiles, C. (1964). The applicability of the third integral of motion: Some numerical experiments, *Astrophys. J.* **69**, pp. 73–79.

Hirsch, M. W. (1988). Systems of differential equations which are competitive or cooperative III: Competing species, *Nonlinearity* **1**, pp. 51–71.

Hirsch, M. W., Smale, S., and Devaney, R. L. (2004). *Dynamical Systems, and an Introduction to Chaos* (2nd edn) (Elsevier/Academic Press, Amsterdam).

Hodgkin, A. and Huxley, A. (1952). A quantitative description of membrane current and its application to conduction and excitatiion in nerve, *J. Physiol.* **117**, pp. 500–544.

Hoover, W. G. (1995). Remark on 'Some simple chaotic flows,' *Phys. Rev. E* **51**, pp. 759–760.

Horowitz, P. and Hill, W. (1989). *The Art of Electronics* (2nd edn) (Cambridge University Press, Cambridge).

Horton, W., Weigel, R. S., and Sprott, J. C. (2001). Chaos and the limits of predictability for the solar-wind-driven magnetosphere-ionosphere system, *Phys. Plasmas* **8**, pp. 2946–2952.

Hoshi, M. and Kono, M. (1988). Rikitake two-disk dynamo system: Statistical properties and growth of instability, *J. Geophys. Res.* **93**, pp. 11643–11654.

Hutchinson, G. E. (1948). Circular causal mechanisms in ecology, *Ann. New York Acad. Sci.* **50**, pp. 221–246.

Ikeda, K. (1979). Multiple-valued stationary state and its instability of the transmitted light by a ring cavity system, *J. Opt. Comm.* **30**, pp. 257–261.

Ikeda, K. and Matsumoto, K. (1987). High-dimensional chaotic behavior in systems with time-delayed feedback, *Phys. Nonlinear Phenom.* **29**, pp. 223–235.

Innocenti, G., Genesio, R., and Ghilardi, C. (2008). Oscillations and chaos in simple quadratic systems, *Chaos* **18**, pp. 1917–1937.

Ito, K. (1980). Chaos in the Rikitake two-disc dynamo system, *Earth Planet. Sci. Lett.* **51**, pp. 451–456.

Jackson, E. A. and Kodogeorgiou, A. (1992). A coupled Lorenz-cell model of Rayleigh-Bénard turbulence, *Phys. Lett. A* **168**, pp. 270–275.

Jia, Q. (2007). Hyperchaos generated from the Lorenz chaotic system and its control, *Phys. Lett. A* **366**, pp. 217–222.

Josić, K. and Wayne, C. E. (2000). Dynamics of a ring of diffusively coupled Lorenz oscillators, *J. Stat. Phys.* **98**, pp. 1–30.

Kapitaniak, T. and Chua, L. O. (1994). Hyperchaotic attractors of unidirectionally-coupled Chua's circuits, *Int. J. Bifurcat. Chaos Appl. Sci. Eng.* **4**, pp. 477–482.

Kaplan, J. and Yorke, J. (1979). Chaotic behavior of multidimensional difference equations. In *Functional Differential Equations and Approximation of Fixed Points, Lecture Notes in Mathematics*, Vol. 730 (ed. H. -O. Peitgen and H. -O. Walther), pp. 228–237 (Springer, Berlin).

Keller, J. M. (1942). Newton's third law and electrodynamics, *Am. J. Phys.* **10**, pp. 1005–1017.

Kennedy, M. P. (1992). Robust op amp realization of Chua's circuit, *Frequenz* **46**, pp. 66–80.

Kevrekidis, I. G., Nicolaenko, B., and Scovel, J. C. (1990). Back in the saddle again: A computer assisted study of the Kuramoto-Sivashinsky equation, *SIAM J. Appl. Math.* **50**, pp. 760–790.

Khibnik, A. I., Roose, D., and Chua, L. O. (1993). On periodic orbits and homoclinic bifurcations in Chua's circuit with a smooth nonlinearity, *Int. J. Bifurcat. Chaos Appl. Sci. Eng.* **3**, pp. 363–384.

Kiers, K., Klein, T., Kolb, J., Price, S., and Sprott, J. C. (2004a). Chaos in a nonlinear analog computer, *Int. J. Bifurcat. Chaos Appl. Sci. Eng.* **14**, pp. 2867–2873.

Kiers, K., Schmidt, D., and Sprott, J. C. (2004b). Precision measurements of a simple chaotic circuit, *Am. J. Phys.* **72**, pp. 503–509.

Kiliç, R. and Yildirim, F. (2008). A survey of Wien bridge-based chaotic oscillators: Design and experimental issues, *Chaos, Solitons & Fractals* **38**, pp. 1394–1410.

Kitchens, B. P. (1998). *Symbolic Dynamics: One-sided, Two-sided and Countable State Markov Shifts* (Springer Universitext, Berlin).

Kolmanovski, V. and Myshkis, A. (1999). *Introduction to the Theory and Applications of Functional Differential Equations* (Kluwer Academic Publishers, Dordrecht).

Korteweg, D. J. and de Vries, G. (1895). On the change of long waves advancing in a rectangular canal, and on a new type of long stationary waves, *Phil. Mag.* (Series 5) **39**, pp. 422–433.

Krogdahl, W. S. (1955). Stellar pulsation as a limit-cycle phenomenon, *Astrophys. J.* **122**, pp. 43–51.

Kuang, Y. (1993). *Delay Differential Equations with Applications in Population Dynamics* (Academic Press, San Diego).

Kudryashov, N. A. (1990). Exact solutions of the generalized Kuramoto-Sivashinsky equation, *Phys. Lett. A* **147**, pp. 287–291.

Kuramoto, Y. and Tsuzuki, T. (1976). Persistent propagation of concentration waves in dissipative media far from thermal equilibium, *Progr. Theor. Phys.* **55**, pp. 356–369.

Kuznetsov, Y. A. (1995). *Elements of Applied Bifurcation Theory* (2nd edn) (Springer, New York).

Lainscsek, C., Lettellier, C., and Gorodnitsky, I. (2003). Global modeling of the Rössler system from the z-variable, *Phys. Lett. A* **314**, pp. 409–427.

Laje, R., Gardner, T., and Mindlin, G. B. (2001). Continuous model for vocal fold oscillations to study the effect of feedback, *Phys. Rev. E* **64**, pp. 056201-1–7

Laskar, J. (1989). A numerical experiment on the chaotic behaviour of the Solar System, *Nature* **338**, pp. 237–238.

Lax, P. and Levermore, C. D. (1983). The small dispersion limit of the Korteweg-de Vries equation, Part I, II, III, *Comm. Pure App. Math.* **36**, pp. 253–290, 571–593, 809–830.

Leipnik, R. B. and Newton, T. A. (1979). Double strange attractors in rigid body motion with linear feedback control, *Phys. Lett. A* **86**, pp. 63–67.

Letellier, C. and Valée, O. (2003). Analytical results and feedback circuit analysis for simple chaotic flows, *J. Phys. Math. Gen.* **36**, pp. 11229–11245.

Levi, M. (1981). Qualitative analysis of the periodically forced relaxation oscillations, *Memoir. Am. Math. Soc.* **214**, pp. 1–147.

Levinson, N. (1949). A second order differential equation with singular solutions, *Ann. Math.* **50**, pp. 127–153.

Li, Y. C. (2004). *Chaos in Partial Differential Equations* (International Press, Somerville, MA).

Li, T. -Y. and Yorke, J. A. (1975). Period three implies chaos, *Am. Math. Mon.* **82**, pp. 985–992.

Li, Y., Tang, W. K. S., Chen, G. (2005a). Generating hyperchaos via state feedback control, *Int. J. Bifurcat. Chaos Appl. Sci. Eng.* **15**, pp. 3367–3375.

Li, Y., Tang, W. K. S., Chen, G. (2005b). Hyperchaos evolved from the generalized Lorenz equation, *Int. J. Circ. Theor. Appl.* **33**, pp. 235–251.

Lichtenberg, A. J. and Lieberman, M. A. (1992). *Regular and Chaotic Dynamics* (2nd edn), Applied Mathematical Sciences, Vol. **38** (Springer, New York).

Lin, C. D. (1995). Hyperspherical coordinate approach to atomic and other Coulombic three-body systems, *Phys. Rep.* **257**, pp. 1–83.

Lind, D. and Marcus, B. (1995). *An Introduction to Symbolic Dynamics and Coding* (Cambridge University Press, Cambridge).

Linsay, P. S. (1981). Period doubling and chaotic behavior in a driven anharmonic oscillator, *Phys. Rev. Lett.* **19**, pp. 1349–1352.

Linz, S. J. (1997). Nonlinear dynamical models and jerky motion, *Am. J. Phys.* **65**, pp. 523–526.

Linz, S. J. (2000). No-chaos criteria for certain jerky dynamics, *Phys. Lett. A* **275**, pp. 204–210.

Linz, S. J. (2008). On hyperjerky systems, *Chaos, Solitons & Fractals* **37**, pp. 741–747.

Linz, S. J. and Sprott, J. C. (1999). Elementary chaotic flow, *Phys. Lett. A* **259**, pp. 240–245.

Liu, W. and Chen, G. (2003). A new chaotic system and its generation, *Int. J. Bifurcat. Chaos Appl. Sci. Eng.* **13**, pp. 261–267.

Liu, Z. Lai, Y. C., and Matías, M. A. (2003). Universal scaling of Lyapunov exponents in coupled chaotic oscillators, *Phys. Rev. E* **67**, pp. 045203(R)-1–4.

Lorenz, E. N. (1963). Deterministic nonperiodic flow, *J. Atmos. Sci.* **20**, pp. 130–141.

Lorenz, E. N. (1991). Dimension of weather and climate attractors, *Nature* **353**, pp. 241–244.

Lorenz, E. N. (1993). *The Essence of Chaos* (University of Washington Press, Seattle).

Lorenz, E. N. and Emanuel, K. A. (1998). Optimal sites for supplementary weather observations: Simulation with a small model, *J. Atmos. Sci.* **55**, pp. 399–414.

Losson, J., Mackey, M. C., and Longtin, A. (1993). Solution multistability in first-order nonlinear differential delay equations, *Chaos* **3**, pp. 167–176.

Lotka, A. J. (1925). *Elements of Physical Biology* (Williams and Wilkins, Baltimore).

Low, L. A., Reinhall, P. G., and Storti, D. W. (2003). An investigation of coupled van der Pol oscillators, *J. Vib. Acoust.* **125**, pp. 162–169.

Lü, J. and Chen, G. (2006). Generating multiscroll chaotic attractors: Theories, methods and applications, *Int. J. Bifurcat. Chaos Appl. Sci. Eng.* **16**, pp. 775–858.

Lu, H. and He, Z. (1996). Chaotic behavior in first-order autonomous continuous-time systems with delay, *IEEE Trans. Circ. Syst.* **43**, pp. 700–702.

Lu, H., He, Y., and He, Z. (1998). A chaos-generator: Analyses of complex dynamics of a cell equation in delayed cellular neural networks, *IEEE Trans. Circ. Syst.* **45**, pp. 178–181.

Lü, J., Chen, G., Cheng, D., and Celikovsky, S. (2002). Bridge the gap between the Lorenz system and the Chen system, *Int. J. Bifurcat. Chaos Appl. Sci. Eng.* **12**, pp. 2917–2926.

Maartens, R., Lesame, W. M., and Ellis, G. F. R. (1998). Newtonian-like and anti-Newtonian universes, *Classical Quant. Grav.* **15**, pp. 1005–1017.

Mackey, M. C. and Glass, L. (1977). Oscillation and chaos in physiological control systems, *Science* **197**, pp. 287–289.

Mahla, A. I. and Badan Palhares, A. G. (1993). Chua's circuit with a discontinuous nonlinearity, *J. Circ. Syst. Comput.* **3**, pp. 231–237.

Mahmoud, G. M. and Bountis, T. (2004). The dynamics of systems of complex nonlinear oscillators: A review, *Int. J. Bifurcat. Chaos Appl. Sci. Eng.* **14**, pp. 3821–3846.

Malasoma, J. -M. (2000). What is the simplest dissipative chaotic jerk equation which is parity invariant?, *Phys. Lett. A* **264**, pp. 383–389.

Malasoma, J. -M. (2002). A new class of minimal chaotic flows, *Phys. Lett. A* **305**, pp. 52–58.

Mandelbrot, B. B. (1983). *The Fractal Geometry of Nature* (Freeman, New York).

Marchal, C. (1990). *Three-Body Problem* (Elsevier, Amsterdam)

Marcus, C. M. and Westervelt, R. M. (1989). Stability of analog neural networks with delay, *Phys. Rev. A* **39**, pp. 347–349.

Marshall, D. and Sprott, J. C. (2009) Simple driven chaotic oscillators with complex variables, *Chaos* **19**, pp. 013124-1–7.

Matsumoto, T., Chua, L. O., and Komuro, M. (1985). The double scroll, *IEEE Trans. Circ. Syst.* **33**, pp. 797–818.

May, R. M. (1975). *Stability and Complexity in Model Ecosystems* (2nd edn) (Princeton University Press, Princeton, NJ).

McArthur, R. H. (1970). Species packing and competitive equilibrium for many species, *Theor. Popul. Biol.* **1**, pp. 1–11.

Moffat, H. K. (1979). A self-consistent treatment of simple dynamo systems, *Geophys. Astrophys. Fluid Dynam.* **14**, pp. 147–166.

Moon, F. C. and Holmes, W. T. (1979). A magnetoelastic strange attractor, *J. Sound Vib.* **65**, pp. 275–296.

Moore, D. W. and Spiegel, E. A. (1966). A thermally excited non-linear oscillator, *Astrophys. J.* **143**, pp. 871–887.

Morgül, Ö (1995). Inductorless realisation of chua oscillator, *Electron. Lett.* **31**, pp. 1303–1304.

Munmuangsaen, B. and Srisuchinwong, B. (2009). A new five-term simple chaotic attractor, *Phys. Lett. A* **373**, pp. 4038–4043.

Murali, K., Lakshmanan, M., and Chua, L. O. (1994). Bifurcation and chaos in the simplest dissipative non-autonomous circuit, *Int. J. Bifur. Chaos Appl. Sci. Eng.* **4**, pp. 1511–1524.

Nagumo, J., Arimoto, S., and Yoshizawa, S. (1962). An active pulse transmission line simulating nerve axon, *Proc. Inst. Radio Eng.* **50**, pp. 2061–2070.

Namajūnas, A. and Tamaševičius, A. (1995). Modified Wien-bridge oscillator for chaos, *Electron. Lett.* **31**, pp. 335–336.

Namajūnas, A., Pyragas, K., and Tamaševičius, A. (1995). An electronic analog of the Mackey-Glass system, *Phys. Lett. A* **201**, pp. 42–46.

Nelson, P. W. and Perelson, A. S. (2002). Mathematical analysis of delay differential equation models of HIV-1 infection, *Math. Biosci.* **179**, pp. 73–94.

Ning, C. and Haken, H. (1990). Detuned lasers and the complex Lorenz equations: Subcritical and supercritical Hopf bifurcations, *Phys. Rev. A* **41**, pp. 3826–3837.

Nishio, Y., Inaba, N., Mori, S., and Saito, T. (1990). Rigorous analysis of windows in a symmetric circuit, *IEEE Trans. Circ. Syst.* **37**, pp. 473–487.

Nosé, S. (1991). Constant temperature molecular dynamics methods, *Prog. Theor. Phys. Suppl.* **103**, pp. 1–46.

Özoğuz, S., Elwakil, A. S., and Salama, K. N. (2002). *n*-scroll chaos generator using nonlinear transconductors, *Electron. Lett.* **38**, pp. 685–686.

Passos, D. and Lopes, I. (2008). Phase space analysis: The equilibrium of the solar magnetic cycle, *Solar Phys.* **250**, pp. 403–410.

Pastor, I., Pérez-García, V. M., Encinas-Sanz, F., and Guerra, J. M. (1993). Ordered and chaotic behavior of two van der Pol oscillators, *Phys. Rev. E* **48**, pp. 171–182.

Pastor-Díaz, I and Lopez-Fraguas, A. (1995). Dynamics of two coupled van der Pol oscillators, *Phys. Rev. E* **52**, pp. 1480–1489.

Patidar, V. and Sud, K. K. (2005). Bifurcation and chaos in simple jerk dynamical systems, *Pramana – Journal of Physics* **64**, pp. 75–93.

Pauli, W. (1922). Über das modell des wasserstoffmolekülions, *Ann. Phys.* **68**, pp. 177–240.

Pesin, Y. B. and Sinai, Y. G. (1991). Space-time chaos in chains of weakly hyperbolic mappings, *Advances in Soviet Math.* **3**, pp. 165–168.

Peters, R. D. and Pritchett, T. (1997). The not-so-simple harmonic oscillator, *Am. J. Phys.* **65**, pp. 1067–1073.

Poincaré, H. (1890). Sur le problème des trois corps et les équations de la dynamique, *Acta. Math.* **13**, pp. 1–270.

Pomeau, Y. and Manneville, P. (1980). Intermittent transition to turbulence in dissipative dynamical systems, *Comm. Math. Phys.* **74**, pp. 189–197.

Press, W. H., Teukolsky, S. A., Vetterling, W. T., and Flannery, B. P. (2007). *Numerical Recipes: The Art of Scientific Computing* (3rd edn) (Cambridge University Press, Cambridge).

Qi, G., Du, S., Chen, G., Chen, Z., and Yuan, Z. (2005). On a four dimensional chaotic system, *Chaos, Solitons & Fractals* **23**, pp. 1671–1682.

Qi, G., van Wyk, M. A., van Wyk, B. J., and Chen, G. (2008). On a new hyperchaotic system, *Phys. Lett. A* **372**, pp. 124–136.

Rand, R. H. and Holmes, P. J. (1980). Bifurcation of periodic motions in two weakly coupled van der Pol oscillators, *Int. J. Non. Lin. Mech.* **15**, pp. 387–399.

Richards, R. (1999). The subtle attraction: Beauty as a force in awareness, creativity, and survival. In *Affect, Creative Experience, and Psychological Adjustment* (ed. S. W. Russ), pp. 195–219 (Brunner/Mazel, Philadelphia).

Rikitake, T. (1958). Oscillations of a system of disk dynamos, *Math. Proc. Camb. Phil. Soc.* **54**, pp. 89–105.

Rollins, R. W. and Hunt, E. R. (1982). Exactly solvable model of a physical system exhibiting universal chaotic behavior, *Phys. Rev. Lett.* **69**, pp. 1295–1298.

Rössler, O. E. (1976). An equation for continuous chaos, *Phys. Lett. A* **57**, pp. 397–398.

Rössler, O. E. (1979a). Continuous chaos – four prototype equations, *Ann. New York Acad. Sci.* **316**, pp. 376–392.

Rössler, O. E. (1979b). An equation for hyperchaos, *Phys. Lett. A* **71**, pp. 155–157.

Rowat, P. F. and Selverston, A. I. (1993). Modeling the gastric mill central pattern generator of the lobster with a relaxation-oscillator network, *J. Neurophysiol.* **70**, pp. 1030–1053.

Rowlands, G. and Sprott, J. C. (2008). A simple diffusion model showing anomalous scaling, *Phys. Plasmas* **15**, pp. 082308-1–7.

Ruby, L. (1996). Applications of the Mathieu equation, *Am. J. Phys.* **64**, pp. 39–44.

Rucklidge, A. M. (1992). Chaos in models of double convection, *J. Fluid Mech.* **237**, pp. 209–229.

Ruelle, D. (1976). A measure associated with axiom A attractors, *Am. J. Math.* **98**, pp. 619–654.

Russell, J. S. (1834). Notice of the reduction of an anomalous fact in hydrodynamics, and a new law of resistance to the motion of floating bodies, *Rep. Br. Assoc. Adv. Sci.*, pp. 531–544.

Saltzman, B. (1962). Finite amplitude free conection as an initial value problem–I, *J. Atmos. Sci.* **19**, pp. 329–341.

Scheffczyk, C., Parlitz, U., Kurz, T., Knop, W., and Lauterborn, W. (1991). Comparison of bifurcation structures of driven dissipative nonlinear oscillators, *Phys. Rev. E* **43**, pp. 6495–6502.

Schiesser, W. E. (1991). *The Numerical Method of Lines: Integration of Partial Differential Equations* (Academic Press, San Diego).

Schiesser, W. E. and Griffiths, G. W. (2009). *A Compendium of Partial Differential Equation Models: Method of Lines Analysis with Matlab* (Cambridge University Press, Cambridge).

Schot, S. H. (1978). The time rate of change of acceleration, *Am. J. Phys.* **46**, pp. 1090–1094.

Shampine, L. F. and Gordon, M. F. (1975). *Computer Solution of Ordinary Differential Equations. The Initial Value Problem* (W. H. Freeman, San Francisco).

Sharpe, F. R. and Lotka, A. J. (1923). Contribution to the analysis of malaria epidemiology. IV. Incubation lag, *Am. J. Hyg.* (Supp. 1) **3**, pp. 96–112.

Shaw, R. (1981). Strange attractors, chaotic behavior, and information flow, *Z. Naturforsch.* **36A**, pp. 80–112.

Shaw, R. (1984). *The Dripping Faucet as a Model Chaotic System* (Aerial Press, Santa Cruz, CA).

Shimizu, T. and Moroika, N. (1980). On the bifurcation of a symmetric limit cycle to an asymmetric one in a simple model, *Phys. Lett. A* **76**, pp. 201–204.

Sivashinsky, G. I. (1977). Nonlinear analysis of hydrodynamic instability in laminar flames, Part 1. Derivation of basic equations, *Prog. Theor. Phys.* **63**, pp. 1177–1206.

Sivashinsky, G. I. and Michelson, D. M. (1980). On the irregular wavy flow of a liquid film flowing down a vertical plane, *Prog. Theor. Phys.* **63**, pp. 2112–2117.

Smale, S. (1967). Differentiable dynamical systems, *Bull. Am. Math. Soc.* **73**, pp. 747–817.

Smith, L. A. (2007). *Chaos: A Very Short Introduction* (Oxford University Press, Oxford).

Sompolinsky, H., Crisanti, A., and Sommers, H. J. (1988). Chaos in random neural networks, *Phys. Rev. Lett.* **61**, pp. 259–262.

Sparrow, C. (1982). *The Lorenz Equations: Bifurcations, Chaos, and Strange Attractors* (Springer, New York).

Sprott, J. C. (1981). *Introduction to Modern Electronics* (Wiley, New York).

Sprott, J. C. (1994). Some simple chaotic flows, *Phys. Rev. E* **50**, pp. R647–650.

Sprott, J. C. (1997a). Simplest dissipative chaotic flow, *Phys. Lett. A* **228**, pp. 271–274.

Sprott, J. C. (1997b). Some simple chaotic jerk functions, *Am. J. Phys.* **65**, pp. 537–543.

Sprott, J. C. (2000a). A new class of chaotic circuit, *Phys. Lett. A* **266**, pp. 19–23.

Sprott, J. C. (2000b). Simple chaotic systems and circuits, *Am. J. Phys.* **68**, pp. 758–763.

Sprott, J. C. (2003). *Chaos and Time-Series Analysis* (Oxford University Press, Oxford).

Sprott, J. C. (2005). Dynamical models of happiness, *Nonlinear Dynam. Psychol. Life Sci.* **9**, pp. 23–36.

Sprott, J. C. (2006). High-dimensional dynamics in the delayed Hénon map, *Electron. J. Theor. Phys.* **3**, pp. 19–35.

Sprott, J. C. (2007a). Maximally complex simple attractors, *Chaos* **17**, pp. 033124-1–6.

Sprott, J. C. (2007b). A simple chaotic delay differential equation, *Phys. Lett. A* **366**, pp. 397–402.

Sprott, J. C. (2008). Chaotic dynamics on large networks, *Chaos* **18**, pp. 023135-1–9.

Sprott, J. C. (2009). Anti-Newtonian dynamics, *Am. J. Phys.* **77**, pp. 783–787.

Sprott, J. C. and Chlouverakis, K. E. (2007). Labyrinth chaos, *Int. J. Bifur. Chaos Appl. Sci. Eng.* **17**, pp. 2097–2108.

Sprott, J. C., Vano, J. A., Wildenberg, J. C., Anderson, M. B., and Noel, J. K. (2005a). Coexistence and chaos in complex ecologies, *Phys. Lett. A* **335**, pp. 207–212.

Sprott, J. C., Wildenberg, J. C., and Vano, J. A. (2005b). A simple spatiotemporal chaotic Lotka-Volterra model, *Chaos, Solitons & Fractals* **26**, pp. 1035–1043.

Squire, P. T. (1985). Pendulum damping, *Am. J. Phys.* **54**, pp. 984–991.

Srzednicki, R. and Wójcik, K. (1997). A geometric method for detecting chaotic dynamics, *J. Differ. Equat.* **135**, pp. 66–82.

Stewart, I. (2000). Mathematics: The Lorenz attractor exists, *Nature* **406**, pp. 948–949.

Storti, D. W. and Rand, R. H. (1982). Dynamics of two strongly coupled van der Pol oscillators, *Int. J. Non. Lin. Mech.* **17**, pp. 143–152.

Strogatz, S. H. (1994). *Nonlinear Dynamics and Chaos with Applications to Physics, Biology, Chemistry, and Engineering* (Addison-Wesley-Longman, Reading, MA).

Strutt, J. W. (1883). On maintained vibrations, *Phil. Mag.* **15**, pp. 229–235.

Strutt, J. W. (1887). On the maintenance of vibrations by forces of double frequency, and on the propagation of waves through a medium endowed with periodic structure, *Phil. Mag.* **24**, pp. 145–159.

Suarez, M. J. and Schopf, P. S. (1988). A delayed action oscillator for ENSO, *J. Atmos. Sci.* **45**, pp. 3283–3287.

Sun, K. and Sprott, J. C. (2009). Dynamics of a simplified Lorenz system, *Int. J. Bifurcat. Chaos Appl. Sci. Eng.* **19**, pp. 1357–1366.

Sussman, G. J. and Wisdom, J. (1992). Chaotic evolution of the Solar System, *Science* **257**, pp. 56–62.

Tamaševičius, A., Mykolaitis, G., and Bumelienė, S. (2006). Delayed feedback chaotic oscillator with improved spectral characteristics, *Electron. Lett.* **42**, pp. 736–737.

Tang, K. S., Man, K. F., Zhong, G. Q., and Chen, G. (2001a). Generating chaos via $x|x|$, *IEEE Trans. Circ. Syst.* **48**, pp. 635–641.

Tang, W. K. S., Zhong, G. Q., Chen, G., and Man, K. F. (2001b). Generation of N-scroll attractors via sine function, *IEEE Trans. Circ. Syst.* **48**, pp. 1369–1372.

Tanner, G., Richter, K., and Rost, J. -M. (2000). The theory of two-electron atoms: Between ground state and complete fragmentation, *Rev. Mod. Phys.* **72**, pp. 497–544.

Testa, J., Pérez, J., and Jeffries, C. (1982). Evidence for universal chaotic behavior of a driven nonlinear oscillator, *Phys. Rev. Lett.* **48**, pp. 714–717.

Thamilmaran, K., Lakshmanan, M., and Venkatesan, A. (2004). Hyperchaos in a modified canonical Chua's circuit, *Int. J. Bifurcat. Chaos Appl. Sci. Eng.* **14**, pp. 221–243.

Thomas, R. (1999). Deterministic chaos seen in terms of feedback circuits: Analysis, synthesis, 'labyrnth chaos', *Int. J. Bifurcat. Chaos Appl. Sci. Eng.* **9**, pp. 1889–1905.

Thomas, R., Basios, V., Eiswirth, M., Kreul, T., and Rössler, O. E. (2004). Hyperchaos of arbitrary order generated by a single feedback circuit, and the emergence of chaotic walks, *Chaos* **14**, pp. 669–674.

Tôrres, L. A. B. and Aguirre, L. A. (2000). Inductorless Chua's circuit, *Electron. Lett.* **36**, pp. 1915–1916.

Tsuneda, A. (2005). A gallery of attractors from smooth Chua's equation, *Int. J. Bifurcat. Chaos Appl. Sci. Eng.* **15**, pp. 1–50.

Tucker, W. (1999). The Lorenz attractor exists, *Compt. Rendus Acad. Sci. Math.* **328**, pp. 1197–1202.

Ueda, Y. (1979). Randomly transitional phenomena in the system governed by Duffing's equation, *J. Stat. Phys.* **20**, pp. 181–196.

Ueda, Y. (2001). *The Road to Chaos* (2nd edn) (Aerial Press, Santa Cruz, CA).

Vallis, G. K. (1988). Conceptual models of El Niño, *J. Geophys. Res.* **93**, pp. 13979–13991.

Valtonen, M. and Karttunen, H. (2006). *The Three-Body Problem* (Cambridge University Press, Cambridge).

van Buskirk, R. and Jeffries, C. (1985). Observation of chaotic dynamics of coupled nonlinear oscillators, *Phys. Rev. A* **31**, pp. 3332–3357.

van der Pol, B. (1920). A theory of the amplitude of free and forced triode vibrations, *Radio Review* **1**, pp. 701–710, 754–762.

van der Pol, B. (1926). On relaxation oscillations, *Phil. Mag. Ser. 7* **2**, pp. 978–992.

van der Pol, B. and van der Mark, J. (1927). Frequency demultiplication, *Nature* **120**, pp. 363–364.

van der Pol, B. and van der Mark, J. (1928). The heartbeat considered as a relaxation oscillation, and the electrical model of the heart, *Phil. Mag. Ser. 7* **6**, pp. 763–775.

van der Schrier, G. and Maas, L. R. M. (2000). The diffusionless Lorenz equations; Shil'nikov bifurcations and reduction to an explicit map, *Phys. Nonlinear Phenom.* **141**, pp. 19–36.

Vano, J. A., Wildenberg, J. C., Anderson, M. B., Noel, J. K., and Sprott, J. C. (2006). Chaos in low-dimensional Lokta-Volterra models of competition, *Nonlinearity* **19**, pp. 2391–2404.

Verhulst, P. F. (1838). Notice sur la loi que la population poursuit dans son accroissement, *Correspondance Mathématique et Physique* **10**, pp. 113–121.

Villasana, M. and Radunskaya, A. (2003). A delay differential equation model for tumor growth, *J. Math. Biol.* **47**, pp. 270–294.

Virgin, L. N. (2000). *Introduction to Experimental Nonlinear Dynamics: A Case Study in Mechanical Vibration* (Cambridge University Press, Cambridge).

Virk, G. S. (1985). Runge Kutta method for delay-differential systems, *IEE Proc. Contr. Theor. Appl.* **132**, pp. 119–123.

Volterra, V. (1926). Variazioni e fluttuazioni del numero d'individui in specie animali conviventi, *Memorie dell' Accademia dei Lincei* **2**, pp. 31–113.

Wang, L. (2009). 3-scroll and 4-scroll chaotic attractors generated from a new 3-D quadratic autonomous system, *Nonlinear Dynam.* **56**, pp. 453–462.

Wang, Y., Singer, J., and Bau, H. H. (1992). Controlling chaos in a thermal convection loop, *J. Fluid. Mech.* **237**, pp. 479–498.

Wang, X. F., Zhong, G. -Q, Tang, K. -S, Man, K. F., and Liu, Z. -F. (2001). Generating chaos in Chua's circuit via time-delay feedback, *IEEE Trans. Circ. Syst.* **48**, pp. 1151–1156.

Wang, G., Zhang, X., Zheng, Y., and Li, Y. (2006). A new modified hyperchaotic Lü system, *Phys. Stat. Mech. Appl.* **371**, pp. 260–272.

Weinert, K., Webber, O., Husken, M., and Theis, W. (2002). Analysis and prediction of dynamic disturbances of the BTA deep hole drilling process. In *Proc. of the 3rd CIRP International Seminar on Intelligent Computation in Manufacturing Engineering*, (ed. R. Teti), ICME 2002, Ischia, Italy, pp. 297–302.

Weiss, C. O. and Brock, J. (1986). Evidence for Lorenz-type chaos in a laser, *Phys. Rev. Lett.* **57**, pp. 2804–2806.

Wiggens, S. (1988). *Global Bifurcations and Chaos, Analytical Methods* (Springer, New York).

Wildenberg, J. C., Vano, J. A., and Sprott, J. C. (2005). Complex spatiotemporal dynamics in Lotka-Volterra ring systems, *Ecol. Complex.* **3**, pp. 140–147.

Woafo, P. and Kadji, H. G. E. (2004). Synchronized states in a ring of mutually coupled self-sustained electrical oscillators, *Phys. Rev. E* **69**, pp. 046206-1–9.

Wolf, A., Swift, J. B., Swinney, H. L., and Vastano, J. A. (1985). Determining Lyapunov exponents from a time series, *Phys. Nonlinear Phenom.* **16**, pp. 285–317.

Wright, E. M. (1955). A nonlinear difference-differential equation, *J. Reine Angew. Math.* **194**, pp. 66–87.

Yalçin, M. E. and Özoguz, S. (2007). n-scroll chaotic attractors from a first-order time-delay differential equation, *Chaos* **17**, pp. 033112-1–8.

Yalçin, M. E., Suykens, J. A., and Vandewalle, J. C. L. (2005). *Cellular Neural Networks, Multiscroll Chaos and Synchronization* (World Scientific, Singapore).

Yim, G., Ryu, J., and Park, Y. (2004). Chaotic behaviors of operational amplifiers, *Phys. Rev. E* **69**, pp. 045201-1–4.

Zabusky, N. J. and Kruskal, M. D. (1965). Interaction of "solitons" in a collisionless plasma and the recurrence of initial states, *Phys. Rev. Lett.* **15**, pp. 240–243.

Zeghlache, H. and Mandel, P. (1985). Influence of detuning on the properties of laser equations, *J. Opt. Soc. Am. B* **2**, pp. 18–22.

Zhang, F. and Heidel, J. (1997). Non-chaotic behaviour in three-dimensional quadratic systems, *Nonlinearity* **10**, pp. 1289–1303.

Zhang, F. and Heidel, J. (1999). Erratum: Non-chaotic behaviour in three-dimensional quadratic systems, *Nonlinearity* **12**, pp. 739.

Zhezherun, A. (2005). Chaotic behavior in piecewise linear Linz-Sprott equations, *J. Phys. Conf.* **22**, pp. 235–253.

Zhong, G. Q. (1994). Implementation of Chua's circuit with a cubic nonlinearity, *IEEE Trans. Circ. Syst.* **41**, pp. 934–941.

Zhou, W., Xu, Y., Lu, H., and Pan, L. (2008). On dynamics analysis of a new chaotic attractor, *Phys. Lett. A* **372**, pp. 5773–5777.

Index

anomalous diffusion, 107
anti-Newtonian system, 142
antidamping, 43
antisymmetric DDE, 229
asymmetric DDE, 229
asymmetric logistic DDE, 230
attractor, 8
attractor dimension, 29
autonomous, 11
autonomous relaxation oscillator, 237

basin of attraction, 32
bifurcation, 22
Bonhoeffer–van der Pol model, 124
boundary condition, 196
breakover diode, 235

carrying simplex, 159
celestial motion, 2
center, 7
chaos, 10
chaotic attractor, 12
chaotic circuit, 233
chaotic sea, 16, 96
chaotic transient, 31
Chlouverakis system, 148
Chua's circuit, 246
Chua's diode, 49
Chua's system, 90
Chua, Leon, 248
circuit elegance, 233
circulant system

conservative, 101
dissipative, 83
high-dimensional, 165
comparator circuit, 256
complex oscillator
autonomous, 156
coupled, 124
forced, 57
ring, 182
star, 191
conjugate pair, 128
conservative chaos, 16
conservative system, 95
continuous delay, 232
continuum ring system, 204
Coulomb damping, 49
Coulomb system, 138
coupled chaotic system, 154
coupled pendulum
circulant ring, 177
circulant star, 187
two identical, 121
coupled relaxation oscillator, 239
crackle system, 147

damping, 8
delay differential equation, 221
delay partial differential equation, 222
delay-line oscillator, 263
delayed-action oscillator, 227
dependent variable, 3
determinism, 1

diac, 235
differential delay equation, 221
diffusionless Lorenz system, 64
dimension, 30, 168
dissipation, 7
dissipative system, 61
Dixon system, 109
drift ring, 16, 97
dripping faucet, 238
Duffing oscillator, 44
Duffing, Georg, 44
dust, 16
dynamical system, 1

Earnshaw's theorem, 139
elegance
    circuit, 233
    equation, 37
ellipsoid, 25
equilibrium, 7
ergodicity, 21
Euler method, 222
Euler's three-body problem, 136
exponential oscillator, 51

fat fractal, 16
feedback, 10
ferrite core, 251
field effect transistor, 252
filamentation, 26
FitzHugh–Nagumo oscillator
    circulant ring, 180
    circulant star, 191
    two coupled, 123
flow, 5
forced diode resonator, 242
forced oscillator, 147
forced pendulum, 116
forced piecewise-linear circuit, 246
forced relaxation oscillator, 234
forced system, 41
forcing, 11
forward Euler method, 222
fractal, 12
frequency demultiplication, 234
friction, 7, 49

functional differential equation, 221

globally coupled network, 187
golden mean, 240
gravitational system, 134
gyrator circuit, 252

Hénon's method, 13
Hénon–Heiles system, 132
Halvorsen's system, 84
Hamilton, William Rowland, 127
Hamiltonian system, 126
high-dimensional system, 115
Hodgkin–Huxley model, 123
Hooke's law, 44
Hutchinson's equation, 227
hyperchaos, 28
hyperchaotic systems, 152
hyperjerk system, 147
hyperlabyrinth system, 173
hyperviscosity, 199
hyperviscous ring, 176
hypervolume, 33
hysteresis, 35, 236, 245, 251

Ikeda DDE, 223
independent variable, 3
intermittency, 32
invariant circle, 9
iron core, 243

jerk circuit, 254
jerk system
    circulant ring, 185
    circulant star, 194
    conservative, 98
    dissipative, 70

Kaplan–Yorke dimension, 30
Kepler, Johannes, 135
Korteweg–de Vries equation, 205
Kuramoto–Sivashinsky equation, 197
Kuramoto–Sivashinsky variant, 200

labyrinth chaos, 105

Lagrange interpolating polynomial, 196
Lagrange, Joseph Louis, 127
Landauer, Rolf, 239
Laplacian, 183
Larsen effect, 10
Leibniz, Gottfried, 4
libration, 5
limit cycle, 8
Liouville's theorem, 7, 128
logistic DDE, 227
Lorenz system
  classic, 61
  coupled, 182
  diffusionless, 64
  diffusionless coupled ring, 185
  diffusionless coupled star, 193
  diffusively coupled ring, 183
  viscously coupled ring, 183
Lorenz, Edward, vii, 61
Lorenz–Emanuel system, 165
Lotka–Volterra system
  3-dimensional, 88
  4-dimensional, 157
  circulant ring, 169
low-dimensional system, 109
Lyapunov exponent
  largest, 20
  spectrum, 24, 117
Lyapunov, Aleksandr, 20

Mackey–Glass equation, 223
magnetic flux, 244
Malasoma system
  cubic, 77
  quadratic, 76
manifold, 7
master–slave oscillator, 259
master–slave system, 116
Mathieu's equation, 55
mean field approximation, 187
memory oscillator, 82
method of lines, 197
model
  AC: autonomous complex, 157
  CJ: conservative jerk, 101
  CO: coupled oscillator, 126
  CS: Chlouverakis snap, 149
  CV: Chua variant, 91
  DV: Dixon variant, 111
  FC: forced conservative, 52
  FO: forced oscillator, 118
  FQ: forced damped quadratic, 48
  FZ: forced complex, 57
  JD: simple jerk, 72
  MO: memory oscillator, 74
  MS: master–slave oscillator, 120
  PD: chaotic PDE, 207
  PO: parametric oscillator, 55
  SQ: simple quadratic, 70
  TW: traveling wave PDE, 213
  VF: velocity forced, 53
Moore–Spiegel system, 77
multiscroll system, 87
multistability, 35
mutually coupled oscillator, 120

N-body system, 134
negative resistance, 246
neon lamp, 234
neural network
  circulant, 174
  description, 159
  minimal, 161
Newton's laws, 2, 83, 142
Newton, Isaac, 4, 135
Nishio's circuit, 249
nonautonomous, 11
nonlinear wave, 205
nonlinearity, 27
Nosé–Hoover oscillator, 95
numerical instability, 197

O-point, 7

pandemonium, 31
parameter, 3
parametric oscillator
  coupled, 130
  forced, 55
parasitic circuit, 233
partial differential equation, 195

Elegant Chaos

pendulum
  conservative, 16
  coupled, 121
  dissipative, 7
  equilibrium, 7
  forced, 11, 116
  frictionless, 4
  ring, 177
  simple, 2
  star, 187
periodic window, 22
periodically forced system
  high-dimensional, 115
  low-dimensional, 41
phase angle, 11
phase space, 4
piecewise-linear system
  circulant conservative, 107
  circulant dissipative, 86
  delay differential equation, 229
  electrical circuit, 254
  forced, 48, 246
  simplest dissipative, 80
Planck's constant, 4, 128
planetary motion, 2, 135
plasma, 139
Poincaré section, 12, 97
Poincaré, Henri, 135
Poincaré–Bendixson theorem, 10, 109
point attractor, 8
polynomial DDE, 225
pop system, 147
Prandtl number, 62
predator–prey problem, 142

quadratic oscillator, 47
quantum mechanics, 128
quasiperiodicity, 16, 20

Rössler hyperchaos, 153
Rössler prototype-4 system, 68
Rössler system, 66
Rössler, Otto, 66
random walk, 105, 145
rational jerk, 76
Rayleigh number, 62

Rayleigh oscillator, 43
Rayleigh, Lord, 55
reactance, 16
rectifier circuit, 254
recurrence, 27
relaxation oscillator, 41, 235
repellor, 9
restricted three-body problem, 136
retarded delay differential equation, 221
retarded functional differential equation, 221
reversibility, 7, 96
Rikitake dynamo, 92
ring of oscillators
  electrical circuit, 261
  ODE, 176
ring system, 165
rotation, 6
Runge–Kutta algorithm, 36
Russell, John Scott, 207

saddle point, 7
saturating inductor circuit, 243
sensitive dependence, 20
separatrix, 6
sigmoid, 49, 159
sigmoidal DDE, 227
signum DDE, 227
signum oscillator
  circulant ring, 178
  circulant star, 188
  electrical circuit, 256
  forced, 49
singularity chaos, 110
sink, 8
sinusoidal DDE, 225
skin effect, 251
snap hyperchaos, 154
snap system, 147
softening spring, 45
soliton, 205
source, 9
spatial chaos, 196
spatiotemporal system, 195
spectral method, 199

Spiegel, Edward, 31, 77
Sprott system, 68
stability, 7
stable manifold, 7
star system, 185
state space, 2
stiffening spring, 45
stochasticity, 1
strange attractor, 10
streamline, 5
symbolic dynamic, 145

temporal chaos, 196
Thomas' system, 85
Thomas, René, 85
three-body problem, 136
time-delay system, 221
torus, 16, 96
transient chaos, 31
traveling wave, 201
traveling wave variant, 212
turbulence, 170, 183

Ueda, Yoshisuke, 46
ultraviolet catastrophe, 173

unidirectional coupling, 116
unstable manifold, 7

van der Pol oscillator
  autonomous, 9
  coupled, 123
  forced, 19, 41
  ring, 179
  star, 190
van der Pol, Balthasar, 41, 234
variable, 2
varicap, 242
velocity coupled oscillator, 129
velocity forcing, 53
Verhulst equation, 230

wave equation, 212
Wien-bridge oscillator, 251
worm hole, 105
Wright's equation, 227

X-point, 7

Zener diode, 250

Printed in the United States
By Bookmasters